D0383068

Bromeliads

BROMELIADS

JACK KRAMER

HARPER & ROW, PUBLISHERS, NEW YORK
CAMBRIDGE, PHILADELPHIA, SAN FRANCISCO, LONDON, MEXICO CITY, SÃO PAULO, SYDNEY
1817

OTHER BOOKS BY JACK KRAMER:

1,000 BEAUTIFUL HOUSE PLANTS

ORCHIDS: FLOWERS OF ROMANCE AND MYSTERY

1,000 BEAUTIFUL GARDEN PLANTS

YOUR HOMEMADE GREENHOUSE

100 GARDEN PLANS

BROMELIADS: THE COLORFUL HOUSEPLANTS

For information address Harper & Row, Publishers, Inc., 10 East 53rd Street, New York, N.Y. 10022. Published simultaneously in Canada by Fitzhenry & Whiteside Limited, Toronto.

FIRST EDITION

Designed by Karolina Harris
Photographs by C. and D. Luckhart except where noted
Drawings by Carol Carlson

Library of Congress Cataloging in Publication Data

Kramer, Jack, 1927–
Bromeliads.

Bibliography: p.
Includes index.
1. Bromeliaceae. I. Title.
SB413.B7K69 1981 635.9′3422 81–47508
ISBN 0–06–038006–3 AACR1

81 82 83 84 85 10 9 8 7 6 5 4 3 2 1

Contents

Preface ix
Acknowledgments xi

PART 1/ CULTIVATING BROMELIADS

One: What You Should Know About Bromeliads 3

Names and More Names 3
Natural Habitat 5
Anatomy of the Bromeliad 8
Blooming Times 13
For Foliage and for Flowers 15
Easy Care 18
Your First Bromeliads 20
Plant Shops, Nurseries, and Mail-Order Suppliers 22
Cost 24
Arrival at Home 25
Collecting Bromeliads 25
Endangered Species 27

Two: Bromeliads at Home 28

Interior Decoration 28
Containers 29
Window Arrangements 33
Mounted Bromeliads 36
Artificial Light 41
How Plants Use Light 42
How Much Light? 43
Fluorescent Lamps 43
Incandescent Lamps 44

Three: Seasonal Culture and Care **45**

Planting Mediums 45
Water and Misting 48
Feeding 48
Light 49
Humidity, Temperature, Air Circulation 49
Grooming and Trimming 49
Potting and Repotting 50
Year-Round Schedule 52
- **Spring 52**
- **Summer 52**
- **Fall 52**
- **Winter 53**

Propagation 53
- **Offshoots 53**
- **Division 55**
- **Seeds 56**

Forcing Flowers 56

Four: Insects and Other Problems **59**

Preventives 60
- **Old-fashioned Remedies 61**
- **Insecticides 61**

Plant Diseases 63
Fungicides 63

Five: Bromeliads Outdoors and in Greenhouses **64**

Gardens 64
Patios and Terraces 66
Returning Plants Indoors 68
Greenhouses 72

PART 2/ A WORLD OF BROMELIADS

Six: Over Two Hundred Bromeliads for You **79**

Abromeitella 79
Acanthostachys 81
Aechmea 81
Ananas 92
Araeococcus 92
Billbergia 93
Bromelia 98
Canistrum 99
Catopsis 100
Cryptanthus 102
Dyckia 107
Fascicularia 109
Guzmania 110
Hechtia 115
Hohenbergia 117
Neoregelia 118
Nidularium 122
Orthophytum 124

Pitcairnia **126**
Portea **126**
Puya **126**
Quesnelia **129**
Ronnbergia **129**
Streptocalyx **130**
Tillandsia **132**
Vriesea **140**
Wittrockia **147**

Seven: Hybrids, Bigenerics, and Collector's Items **151**

Collector's Plants **151**
Bigenerics **152**
Other Bromeliads **153**

PART 3/ CHOOSING BROMELIADS

Easy-to-Grow Bromeliads **157**
Bromeliads for Direct Light **157**
Bromeliads for Diffused Light **158**
Bromeliads for Semi-Shade **158**
Bromeliads at a Glance **159**

Glossary **167**
Bibliography **169**
Appendix A: Note on Bromeliad Societies **171**
Appendix B: Suppliers **173**
Index **175**

A section of color plates follows page 52

Preface

Bromeliads were perhaps first introduced as houseplants in 1693, when Charles Plummier, a French explorer, described plants known as Bromelia and Karatas in his *Plants of the Americas.* Before that we know that Christopher Columbus found the pineapple—a bromeliad—on the island now known as Guadaloupe while on his second voyage in 1492, and brought it to Spain. By the early 1500s, pineapples were the most exotic fruit grown for the tables of royalty. In 1555 Jean de Lery imported pineapples into England for their fruit, and at the end of the sixteenth century pineapples or Ananas were cultivated in the gardens of Versailles and given to the French king as a gift.

After Plummier introduced bromeliads as decorative plants for indoors, the French traveler Feuillée brought specimens back from Peru and Chile in 1709 and in 1730, while another explorer, Demarchais, found plants in Guinea. And from P. Miller's *Gardeners Dictionary,* published in 1785, we find that certain Bromelia species were being grown in English conservatories. Most of the early bromeliads were classified as either Tillandsias or Bromelias and were described by Carl Linneaus in his *Species Plantarum,* published in 1753. Not until 1805 did the plant family get its official name, when Jaume Saint Hillier renamed the group Bromeliaceae.

By the nineteenth century, bromeliads were beginning to be used for indoor decoration, and such publications as Curtis's *Botanical Magazine* (1815) and the *Botanical Register* (1818) gave ample space to bromeliads. Aechmeas and Billbergias were the earliest plants to be so reported.

By 1864, Kew Gardens in England housed more than 100 different bromeliads, but in conditions hardly suitable for survival. Since the plants had come from tropical countries, they were grown in simulated tropical conditions—hot and humid. Most of the bromeliads, however, came from high elevations—a fact not revealed, or perhaps misunderstood, by collectors of the time—and they died by the score. Still, the novelty of growing them persisted. In Charles McIntosh's *Book of the Garden*, published in 1870, there is an entire chapter devoted to growing pineapples as fruit.

Many species were also in cultivation in Berlin in 1857–60, but it was the Belgians who popularized bromeliads as houseplants. Many horticulturists and explorers sponsored by the Belgian government searched for unusual plants, and the first full treatise on bromeliads as indoor plants appeared in *Die Familie der Bromeliaceen* by Joseph Georg Beer, published in 1857. In 1885, Edouard Morren published his *Belgique Horticole* containing colored plates of plants. Notable work was also done with bromeliads in the latter part of the nineteenth century by C. Koch in Germany and Regel in Russia.

In 1935 the plant family acquired increased recognition with the publication of *Bromeliaceae* by Carl Mez, a German botanist. At about the same time the vogue for bromeliads crossed the Atlantic when Mulford Foster of Orlando, Florida, started growing the plants. His many works on the subject furthered bromeliad growing in America, and many plants bear his name, including *Aechmea fosteriana, Dyckia fosteriana,* and *Cryptanthus fosteriana.* This noted collector, grower, and writer founded the Bromeliad Society in 1950.

I myself started working with bromeliads in 1960 when I was growing orchids. I knew that in their native habitat bromeliads grew mainly on trees with orchids. I decided that the conditions I had duplicated in my apartment for orchids would suit bromeliads as well—and they did. For two years I collected the plants and found that the majority of them grew with minimal care, even less care than the ubiquitous philodendrons. In 1965 I published a book about my experiences with bromeliads, *Bromeliads, the Colorful Houseplants,* the first modern handbook published about the plants for the layman in the United States.

It was not until the early 1970s that bromeliads became well known in America. By then, plants had become readily available and hybridizers had started to experiment with them, introducing many new varieties noted for their leaf color, plant shape, or flower color. I added almost 100 species and varieties to my own collection and found that some bromeliads were better than others, some more handsome, some more amenable to untoward conditions. I also learned new ways of growing the plants—new potting mediums—and found various ways of displaying the plants indoors as beautiful accents. All that information is included here.

As of this publication, bromeliads still have not reached their peak of popularity in this country, for it takes time for new plants to be accepted and approved by the public. It took orchids almost 100 years to become popular houseplants; it may take even less time for bromeliads to follow suit.

Jack Kramer
1981

Acknowledgments

It is difficult to thank each and every person who contributes to a book, but several people went beyond and above the general call of help to assist me in this project. I owe special thanks to Chalmers and Dean Luckhart, my photographers, who traveled far and wide to get photographs of bromeliads and then labored long in the darkroom. Herman Pigors of Oak Hill Gardens in Dundee, Illinois, a longtime friend, again contributed much information, and many of the plants I write about came from his nursery. Bill Seaborn of Escondido, California, answered my questions about certain species and helped greatly in suggesting the more popular species that I might include in the book. Larry Cline worked for several weeks checking plant descriptions, and deserves special thanks. And as always to my typist, Judy Smith, my gratitude.

And to the many bromelaid societies in the United States, again my deepest appreciation, especially the Bromelaid Society of Sacramento, California, who let us photograph at their show.

Part 1/CULTIVATING BROMELIADS

One: What You Should Know About Bromeliads

There are about 2,500 known species of bromeliads, and more are introduced into cultivation yearly. Why are bromeliads becoming increasingly popular? Mainly because these plants can almost, with only minimal care, take care of themselves and bring unparalleled color into the home. Bromeliads do not need excessive light to prosper—they survive even in north windows—and most of the plants have their own water reservoirs, which is handy if you forget to water them or you are away from home for a time. Bromeliads will even survive for many months pinned to a curtain or mounted on a wall without any water—such is their will to survive! Finally, bromeliads are generally small to medium in size, making them ideal for apartments.

NAMES AND MORE NAMES

Most hobbyists find the classification of the bromeliad family complicated, but it need not be. Basically the Bromeliaceae family is divided into three subfamilies: Pitcairnioideae, Bromelioideae, and Tillandsioideae. Most bromeliad plant names are in Greek or Latin form, a usage which is international. The Latin or Greek names usually refer to an outstanding characteristic of the plant, such as red leaves, or the person who discovered it, or the geographical place where the plant was first found.

For example, the genus names *Tillandsia* and *Billbergia* honor two Swedish botanists, Tillands and Billberg. The species name *circinnata* means twisted or rolled, referring to the plant's leaves, while *venezuelana* is a species that was discovered in Venezuela.

The Latinized names are always given in italic type, with the genus name capitalized and the species in lower case; after the first mention, the genus may be designated by its capital letter only.

In addition to natural species, there are plants that have been hybridized to produce some effect such as outstanding flowers or unusual leaf color. These hybrids are called varieties. Varieties are denoted by single quotation marks, e.g., *Aechmea* 'Meteor,' which is known for its brilliant flower bracts. However, as work is still going on with the hybrids, the same varieties may bear different names in different regions.

Bromeliads can also be known by unofficial names. *Aechmea fasciata* is generally known as such worldwide, but its common name in some countries is Grecian Urn and in other regions, Silver Urn. Thus, common names are not always trustworthy; only the botanical name should be used when buying any plant.

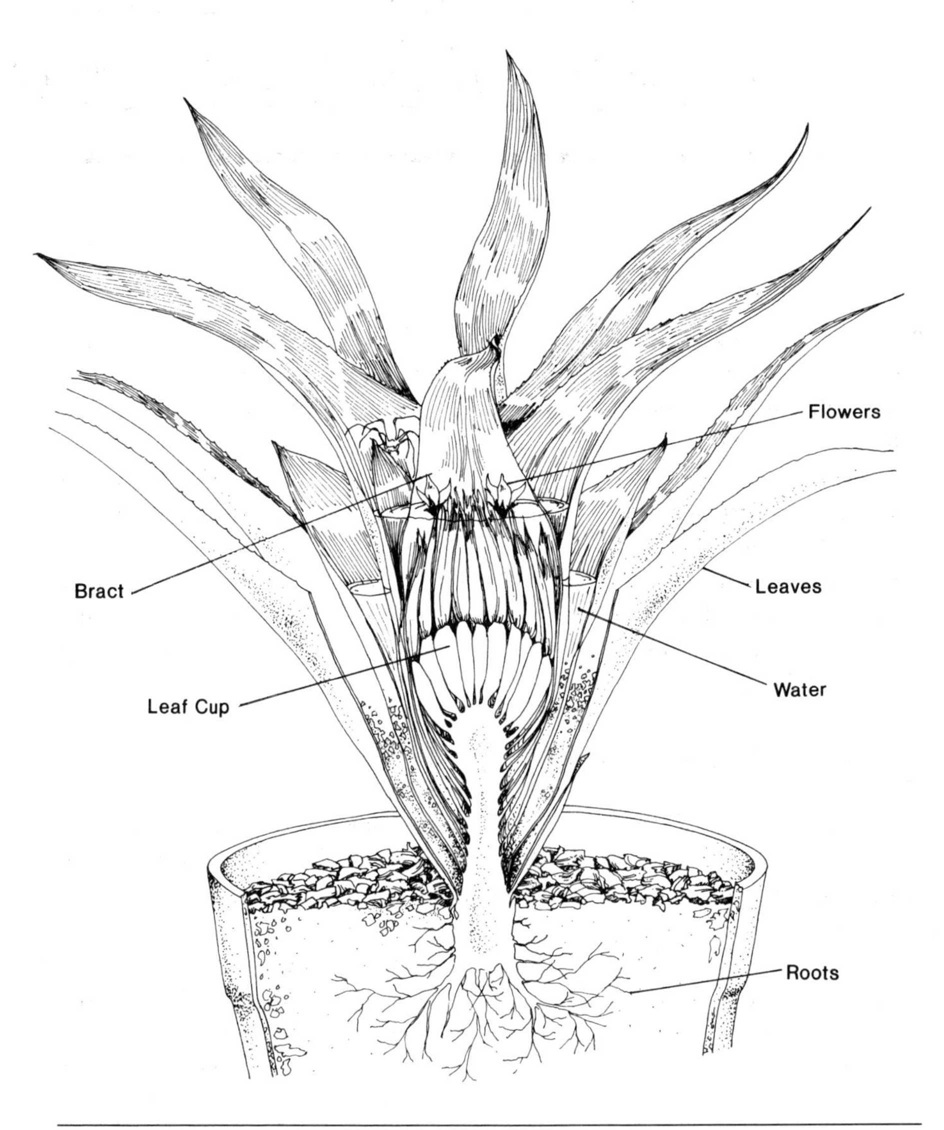

Anatomy of a Bromeliad

NATURAL HABITAT

Bromeliads grow from the state of Virginia southward to southern Argentina, but it is in Mexico, Brazil, Chile, Peru, and Colombia where the largest number of species are found. These lands are literally adorned with bromeliads growing mainly on tree branches or on rocks, some in the ground. Most bromeliads are epiphytic (air-growing) plants clinging to trees and rocks. Their intense coloration can be seen from great distances like tiny jewels sequining the trees.

Most bromeliads, like these large clumps growing in Costa Rica, prefer tree branches to the ground. ***(Photograph courtesy Paul Hutchinson)***

In a South American jungle, bromeliads grow (right) in a dense thicket on trees. Orchids (center) are often found along with them. ***(Photograph courtesy Paul Hutchinson)***

While bromeliads are mainly tropical plants, this is not altogether true. Many do grow in tropical regions, but within these countries plants may be found growing in the plains at low altitudes, in the rain forest at higher heights, and even in the cold regions of the mountains—at about 8,000 feet (2,438 meters) in the Andes. So it is not so much where a plant grows but rather at what elevation it resides that is a clue to its cultural needs, such as temperature and light.

Mexico, rich in bromeliads, has many different "layers" of ecological variation. There are essentially four climatic areas, each with its own group of bromeliads: the hot zone (swamps and wooded regions), the evergreen tropical rain forests, the mountain forests, and finally the cloud forests at about 6,000 feet (1,829 meters) and above. In each individual ecological stratum, there is a definite climatic change—warm to moderately warm, to cool, to quite cool. It is said that for every thousand feet there is typically a temperature variation of from 3°F to 7°F (1.7°C to 3.9°C).

The distribution of plants at varying altitudes is characteristic of the

Puya alpestris, one of the few ground bromeliads, grows in fields in South America. *(Photograph courtesy Dick Lucier)*

Up in the treetops, bromeliads and orchids grow together deep in the rain forest of South America, receiving only dappled sunlight. *(Photograph courtesy Paul Hutchinson)*

South American bromeliad countries as well. In Peru, for example, there are bromeliads growing at five different levels: sand desert, rocky desert, hot, humid rain forests, cloud forests, and in the snow-covered mountains.

Most Neoregelias, Nidulariums, and Guzmanias prefer high treetops or mountaintops (4,000 to 6,000 feet/1,219 to 1,829 meters) because the temperature is warm during the day but quite cool at night (55°F/12.8°C). Other bromeliads, such as the Aechmeas and Billbergias, generally grow on the ground of the forest, where it is somewhat warmer but not overly hot. A few bromeliads grow naturally on other plants or even on telephone wires and other structures. Many species grow under all three conditions.

No matter where they grow, bromeliads almost always inhabit areas where there is good air circulation. Carnivorous plants and some terrestrial philodendrons may grow in steamy, stagnant jungles, but bromeliads thrive only where the air is buoyant because air that brings moisture in the form of dew or fog is essential to their growth.

Sun affects their leaf color, but bromeliads do not need intense sun, and they grow mostly in shaded areas that receive dappled sunlight.

It is not necessary to convert your home into a tropical rain forest or a cloud forest to accommodate bromeliads. In fact, plants grow best under average home temperatures. In most homes, nighttime temperatures are lower than during the day, and this variation is perhaps the most important aspect of general plant care. Bromeliads are notably adaptable plants and, given time, will make themselves at home.

ANATOMY OF THE BROMELIAD

Most bromeliads have a short central trunk, and all bromeliads have scales on their leaves; the scales act as a water-absorbing system. You can see the scales on bromeliads like *Aechmea fasciata*—they form silver bands—but on others you can detect the scales only with a magnifying glass. Leaves differ considerably: some are tough and firm (for example, Aechmeas and Hechtias); several are edged with spines; others, such as the Guzmanias and Vrieseas, have soft, leafy, green foliage. Leaves may be plain green or banded, mottled, streaked, or spotted.

The flower stalks, which in most bromeliads rise from the center of the plants, are tall and wiry; certain plants, like Neoregelias and Nidulariums, have no flower stalks. The flower head is cylindrical, pyramidal, or flat, and it is the color of the head (bract) that makes the plants so outstanding. The flowers are generally minuscule—except for *Tillandsia cyanea*, which has 1-inch (2.5-centimeter) purple blooms. Many bromeliads have brightly colored berries that remain on their stalks for months.

Bromeliads are generally tubular-, vase-, or rosette-shaped—the leaves are curved to form a cup or reservoir for water and food. In the bromeliads' native habitat, small insects and organic matter drop into this reservoir, living and eventually dying there, furnishing nutrients for the plants. In cultivation, bromeliads are free of such pests, although a solitary frog has lived in my *Aechmea fasciata* for more than a year, benefiting from a supply of fresh water and protection from predators! The cup or vase of bromeliads

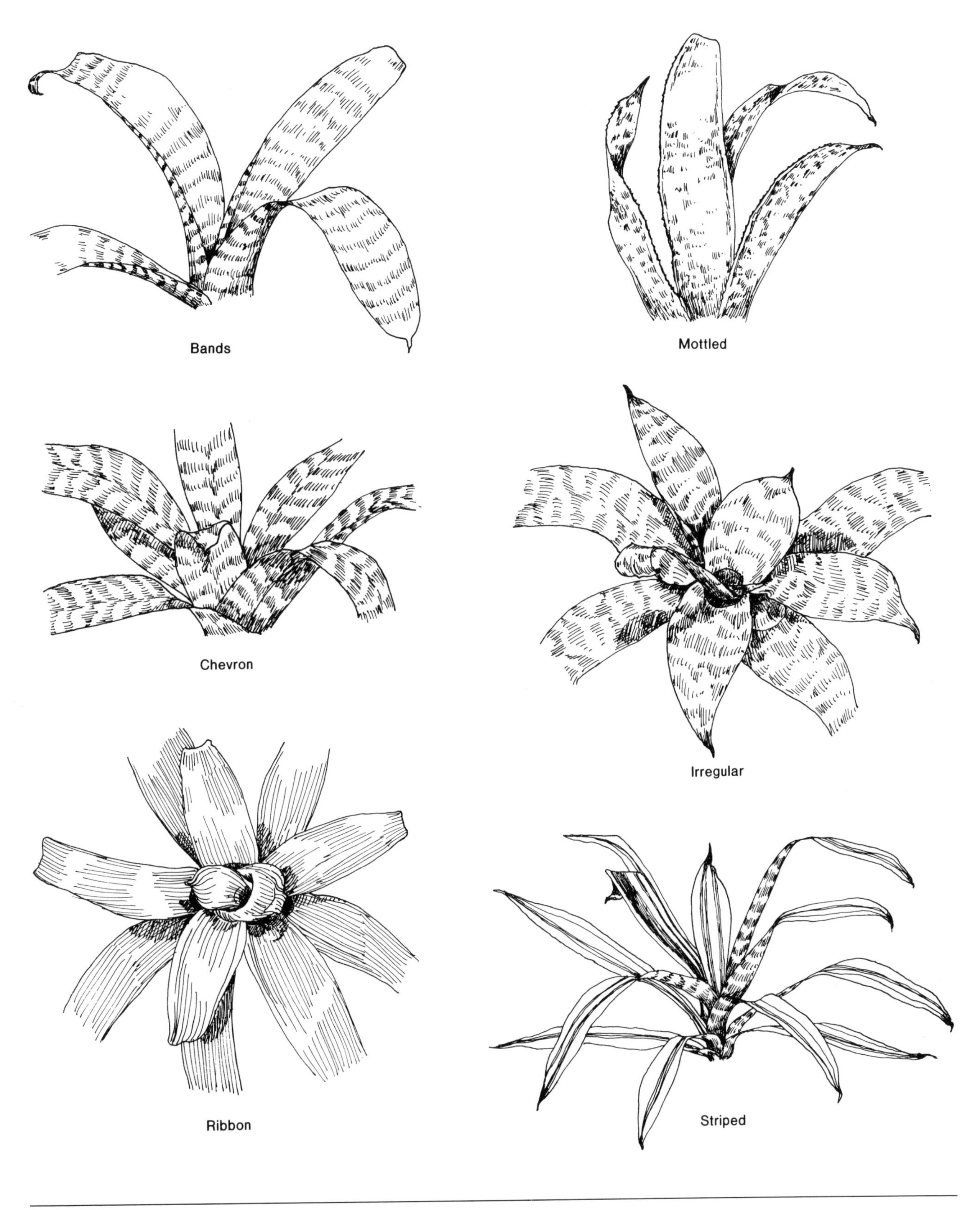

Leaf markings

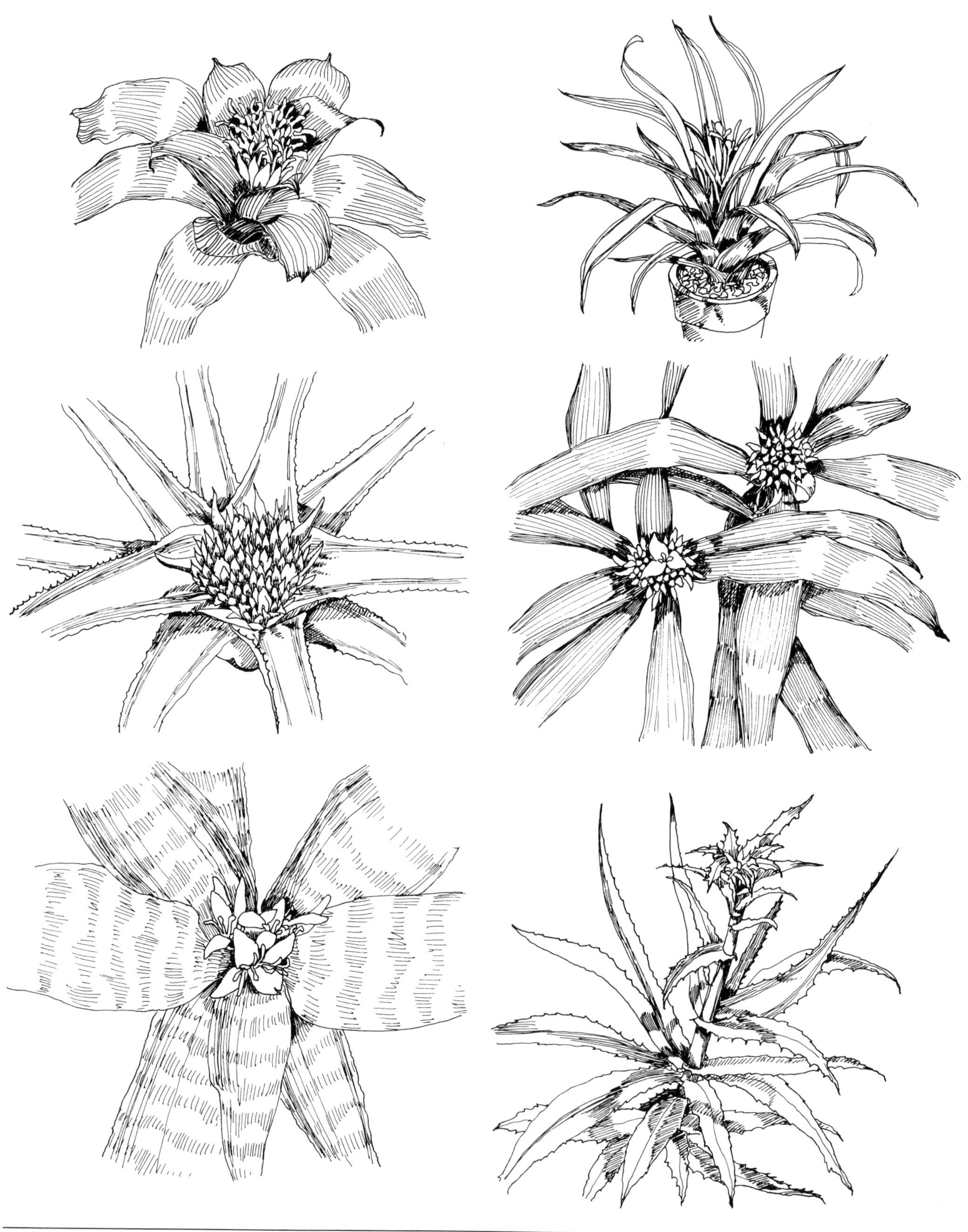

Flower form

Bromeliads excel in leaf patterns; *Aechmea chantinii* 'Burgundy' has dark foliage splashed with lighter color.

Bromeliads at a window show typical growth shapes: tubular at left and right as seen in many Billbergias and Catopsises; center toward window, the rosette type as seen in Guzmanias and Neoregelias; in front, Cryptanthuses in flat rosette pattern. *(Photograph by the author)*

Types of inflorescence

must be kept full of water; the potting medium can be somewhat dry, which will not harm the plant.

Aechmeas have a circular rosette of leaves; Nidulariums and Neoregelias have flattened tops, like a pressed fan. Billbergias have vertical and tubular growth, and Vrieseas have tapered or rosette growth. Dyckia and Cryptanthus species look like starfish because of their spiky growth.

BLOOMING TIMES

Most bromeliads bear their colorful inflorescence in the spring and summer; it is then that my garden room is alive with vibrant colors. Billbergias and Aechmeas predominate, and Neoregelias and Nidulariums offer fiery crowns of color in the late summer.

In the winter, several of my Guzmanias bloom, notably *G. monostachia* and *G. lingulata*, and some Neoregelias are still colorful too. In fact, there are few times during the year when there is not color in my garden room.

Because flower crowns stay colorful so long (up to two months), for most bromeliads it is difficult to specify blooming times as one might do for orchids. (Exceptions are Tillandsias and Billbergias, whose flowers last only a few days.)

Canistrum **foliage is dotted with color and edged with small spines.**

Generally, most bromeliads bloom indoors if there is sufficient light. Some species, like Puyas, Dyckias, and Hechtias, have difficulty blooming because they must have strong sun, which is generally lacking indoors. However, many Vrieseas will bloom with only diffused light—no sun—while Tillandsias require very bright light to bloom. Even without flowers, most bromeliads are worth their space because their foliage color is so decorative.

Vertical lines striating the leaves of a *Vriesea* make a handsome pattern.

FOR FOLIAGE AND FOR FLOWERS

You can grow many species of bromeliads strictly for their highly decorative foliage. For example, the beauty of *Vriesea hieroglyphica* or *Aechmea zebrina* cannot be surpassed. Neoregelias are famous for their leaves, and Cryptanthuses are highly prized for their leaf color. Nidulariums offer beautiful foliage because the symmetrical rosettes are so handsome. An outstanding plant, one I have treasured for years, is *Orthophytum navioides*, with dark-green spiny leaves that are red at the base and shiny like lacquer. *Vriesea fenestralis* and *V. hieroglyphica* have banded, colorful leaves, and not to be slighted are *Guzmania lindenii* and *G. zahnii*.

The branched inflorescence of many bromeliads makes a massive display. The spike may be very tall or somewhat short, as here.

When we speak of flowers in bromeliads, we mean the colorful bracts that retain their beauty for weeks. In this group are the superlative *Aechmea fasciata* and *A. chantinii*, both examples of nature at its best. The former has tufted pink flower heads; the latter, brilliant red and yellow bracts. The Billbergias, specifically *B. pryamidalis* and *B. zebrina*, have very large, vividly colored bracts. The Guzmanias excel in flower crown color, including the red, white, and black *G. monostachia*, the long-lasting orange bracts of *G. lingulata*, and the charming yellow flowers of *G. zahnii*. Streptocalyxes have bizarre but beautiful branched red inflorescence, and some of the Quesnelias look like red-hot poker plants. Tillandsias almost defy description (many of them are discussed in Chapter 6). The brilliant red, candlelike inflorescence of *T. caulescens* and the yellow waxy head of *T. fasciculata* especially deserve mention.

A candelabra type of inflorescence is typical of many bromeliads.

Flower spikes can be pendant, as in *Billbergia nutans*, and stalks may be 3 or 4 feet (91 centimeters to 1.2 meters) long.

The small flowers are hidden within the inflorescence of a *Neoregelia*.

Many bromeliads, like this *Aechmea*, have colorful berries once flowers have faded.

EASY CARE

Bromeliads are easier to care for than, say, orchids or begonias. With few exceptions, bromeliads, because of their growth and makeup, are adaptable plants that can, if necessary, adjust to varying indoor conditions and still survive, whereas ferns and palms, with even the utmost attention, sometimes never become acclimatized indoors.

As with all houseplants, watering, soil, light, humidity, and air circulation are the necessary elements for healthy bromeliads; containers, potting,

A typical bromeliad flower in closeup. Only ¼ inch (.6 centimeter) across with petals folded back, it has funnel-shaped flowers.

The single-head flower crown of *Aechmea fulgens discolor.*

and repotting are other considerations. None of these conditions is difficult to maintain for bromeliads.

We mentioned that bromeliads are more amenable than most plants because their vases act as water reservoirs. Bromeliads' leaves are equally capable of handling varying light conditions. True, the more light a plant has, the more handsome its foliage, but bromeliads, even in a less-than-optimum light situation, can produce attractive foliage.

Most indoor plants rely on extensive root systems to survive, but bromeliads breathe through their leaves and can take in moisture through their foliage, so they will still survive, even if potted improperly. Because most bromeliads are epiphytes or air plants, they can grow without soil altogether if they are misted daily. Many bromeliads can grow in just fir bark or tree fiber chunks, thus eliminating the messy work with soil in the home: no muddy sinks or soil stains on the floor; no sack of soil to carry up to an apartment or into a house. I have a group of six Tillandsias on a piece of tree branch with no soil at all. These plants are in good light, get ample daily misting, and are beautiful.

Since bromeliads can take culture abuse and still survive indoors, they have become a popular houseplant for people who work every day and have little time to care for their plants. Insects are rarely problems for the indoor bromeliads because insects do not care for the plants' tough and leathery leaves. Most bugs generally avoid them and seek easier prey. Mealybugs, a menace to many indoor plants, rarely attack a bromeliad; they migrate to plants that have leafier growth and better places to hide. Aphids too prefer easier prey. However, if your plants get infested with insects, Chapter 4 tells you what to do. And in all my years of growing bromeliads I have yet to see one attacked by the diseases that often decimate houseplants.

Bromeliads are fine houseplants if you follow the general suggestions for cultivation included in this chapter. Chapter 3 contains more detailed information and specific care instructions.

YOUR FIRST BROMELIADS

Bromeliads make any nongardener look good; with them you will always have the pleasure of people telling you what a green thumb you have. However, along with recommending a choice for your first bromeliads, I am also suggesting what you should do with the plants when you get them home—the techniques that can save you trouble and toil later. Much like the baby coming home from the hospital, a new plant needs attention to help it become adjusted to its new environment.

As in any plant family, some species become more popular than others. Within the Bromeliad family, Aechmeas, Guzmanias, and Vrieseas lead the list. Perhaps the best-known Bromeliad is *Aechmea fasciata*. This is a perfectly sized plant, to 28 inches (71 centimeters); it has tubular growth and handsome silvery banded leaves. The flower stalk grows erect from the center of the plant and is a tufted head of pink, with tiny flowers hidden in the bracts. The flowers die quickly, but the flower head lasts for months. This species makes a handsome table or desk accent, and is highly prized. Recent-

The large flower crown of *Aechmea fasciata variegata*, which lasts for many weeks, makes it a popular plant.

ly several hybrids have appeared; they are spectacular indoor plants blazing with colorful foliage and flowers.

Guzmania lingulata is another excellent indoor subject. The plant has shiny apple-green leaves in rosette growth, is small—to 20 inches (50.8 centimeters)—and has a magnificent bright orange flower crown that stays colorful for weeks.

Tillandsia cyanea, native to the forests of Ecuador, has grassy leaves and produces rather large (for the genus) and highly decorative purple flowers. This beautiful plant bears flowers in a shady place, which accounts for its popularity.

Vriesea carinata and *V. splendens* are two robust plants. *V. carinata* is charming, with soft-textured green leaves and golden yellow inflorescence that rises from the center of the plant on a tall scape. The bracts are crimson at the base and yellow at the tips; the flower is bright yellow. *V. splendens*, the flaming sword, is a dull-green rosette of arching leaves with crossbands of purplish black on its underside. It grows to about 20 inches (50.8 centi-

Guzmania lingulata is known for its star-shaped flower head, which varies in color from orange to red.

meters) and the inflorescence is about 16 inches (40.6 centimeters) tall; the lance-shaped spike has orange to red bracts with small yellow flowers.

Neoregelias offer several species that are excellent indoors. Especially favored is *Neoregelia carolinae tricolor*, an exquisite rosette of multicolored leaves, and *N.* 'Marmorata,' with beautifully colored wine-red foliage.

Nidularium innocentii lineatum brings rich green-and-white striped foliage into the home. This spectacular plant is amenable to most conditions.

For limited space, consider several of the fine *Cryptanthus* species. There are varieties with leaf coloring that is perhaps the most beautiful in the plant world.

PLANT SHOPS, NURSERIES, AND MAIL-ORDER SUPPLIERS

In my home state of California there are shops devoted exclusively to bromeliads and orchids. In those I visited, most plants were in soil or on hanging rafts (pieces of bark). If you buy from a plant shop, ask the personnel (1) what the plant is planted in, (2) when it was potted, (3) how old the plant is, and (4) where it comes from.

Also ask for plants by their scientific names; common names can be different in different regions. Generally, plant shops are fine places at which to buy bromeliads because most of the owners are knowledgeable about the plants. Also, the pots are tagged with names; if they are not, ask for the botanical name.

Outdoor nurseries also sell bromeliads. The plant material is satisfactory, but I am leery of buying at these nurseries because I know that their main business is outdoor plants. Their indoor plant sales are less important to them, and so they may not sell quality houseplants.

Shopping personally for your bromeliads at a plant store or nursery means that you can inspect plants and select the best ones. A good plant has brilliant leaf color, perky (not wan) leaves, an overall healthy appearance, and it is free of insects. Inspect the foliage carefully to see if plants are potted in the proper medium and in the proper container, such as a clay pot.

If you can go to a mail-order bromeliad nursery, do so because it is a fine place at which to become acquainted with the plants. The owner is interested in selling the plants, and the variety is usually staggering.

An improved form of *Guzmania lingulata* has a fuller and larger head of flowers.

If you cannot get to a nursery, then order plants from its catalog. Most suppliers have good catalogs, and with improved shipping materials and methods there is little risk of losing plants in transit.

Most plants from mail-order suppliers are shipped bare-root or in pots. If possible, buy plants in pots. This means extra weight and thus more shipping cost, but because bare-root plants have been uprooted from their pots and then sealed in closed boxes for travel, by the time you get them they can be in sad shape. Most plants recover in time, becoming fine specimens, but shipping in a pot alleviates any "pain" for the bromeliads.

COST

Generally, bromeliads are inexpensive; the exceptions are new introductions or rare species or varieties. Be prepared to pay about $10 for a mature bromeliad and $5 for a seedling. If you are a beginner, start with mature plants; later, after you have some growing experience, buy seedlings. If prices are higher than $10 for popular varieties like *Aechmea fasciata* or *Guzmania lingulata,* you are buying from the wrong source. Shop more; look at more catalogs.

***Vriesea carinata* (right), *Aechmea fasciata* (center), and *Aechmea angustifolia* (left), make a handsome group of bromeliads.**

Be especially wary of advertisements for collections of several different kinds of bromeliads for $10 because invariably these are inferior plants a grower wants to get rid of. And be especially cautious of full-page ads promoting a specific plant at a low price: it simply does not make sense to offer a "bargain" yet pay thousands of dollars for a full-page ad.

ARRIVAL AT HOME

When you get your plants home, check them again to see if any insects are present. If plants are bare-root, inspect the roots for insects. Also check mail-order plants upon arrival because insects can travel great distances and still survive. Sometimes plants carry invisible insect eggs rather than mature insects. In any case, soak each plant to the pot's rim in a sink full of water for an hour—any uninvited guests will come to the surface. Then spray plants with a strong jet of water to dislodge any eggs. Finally, polish leaves with a damp cloth. Do not use any leaf-shining preparations because all they do is clog the pores of the leaves.

Do not put new arrivals in direct sun because the abrupt change in light can harm them. Place the plants in a semilight position and water them. In a few days, move the plants to a somewhat brighter place for another five days or so. Finally, move plants to their permanent sunny or bright location. During this time, be sure the plants have good air circulation.

COLLECTING BROMELIADS

On a recent trip to the Caribbean I had the opportunity to collect bromeliads from their native habitat. This is an excellent way of securing difficult-to-find species at inexpensive prices. However, often it is difficult to make a positive identification of collected species, in which case you must seek help from a taxonomist, or specialist in plant classification.

There are two ways of securing plants when you are at the source. You can seek out a local dealer who has already collected the specimens and runs a nursery or an outlet. Or you can hire guides and go into the areas and collect your own plants and then have them packaged and shipped. In either case, you must have an import plant permit from the U.S. Government. The Permit Unit, Plant Quarantine Division, 209 River Street, Hoboken, NJ 07030 will give you a permit at no charge. This department will advise you on how to import plants and, depending upon where you are collecting, will send you suitable plant-permit tags for legal entry of the plants at several ports throughout the United States. You must also have an export plant permit from the country you are visiting.

When you collect plants, remove them carefully from their supports; do not pull them loose. Pry gently with a knife, and then ease each plant from its branch or rock. Remove any dead mother plants and basal leaves, and be sure to shake out tubular rosettes to eliminate water and insects. The heart of a collected plant is the clue to a good specimen: pull very gently on the inner leaves—if they are tight, the plant is healthy. If the leaves come out easily, the plant is worthless.

Once you have collected and cleaned your plants, give them a collection

A closeup of *Nidularium innocentii* shows the delicate tracing of the leaves; at bloom time the center turns fiery red.

number and write notes about where you found the species, the exact elevation, and surrounding plant life. If you have a new species in your collection, it can not be described or catalogued without this information.

If you are collecting your own plants, be sure to bring along plastic wrapping and to obtain suitable packaging and mailing materials: cardboard boxes, newspapers, string or rubber bands, and tape. Trim away dead roots (those that are brown) and dead leaves. Wrap each plant separately, and be sure each plant is totally dry before wrapping it. *Do not mist with water before wrapping.* Insert plants in a thin plastic material like a Baggie. Fold newspapers around the plastic, and tie the package with string or rubber bands. Put each package in a cardboard box and mail it to yourself, or, if you have only a few plants, take them with you. Declare the plants at customs, and show your plant-permit tags.

Plants coming into the United States from foreign countries are subject to inspection and fumigation. The insecticides used are hazardous, and some of the plants you collected or bought away from home will succumb to the fumigation process. However, most of the plants will eventually come back with fresh new growth.

If you buy plants from a supplier in a foreign country, the supplier will do all the wrapping and packaging and make all importation arrangements. However, you will still need the plant-permit tags to export the plants.

When your imported plants arrive home, keep them in a somewhat shady spot for a few days because abrupt exposure to bright light or sun after three or four days of travel in a closed box can harm plants. You can leave the plants unpotted, or pot them with fir bark or osmunda (tree-fern root) into containers. (See Chapter 3 for a discussion of potting.) Do not water plants for a few days; thereafter, bring plants into bright light and apply water to their leaves and cups.

ENDANGERED SPECIES

Many countries legally limit the number of plants that can be exported. This is especially true for orchids. Be sure there are no legal restrictions or limitations on plants before you start collecting them. Check with local officials in the countries you are visiting; laws about plants are different in different regions, so to avoid problems later, be sure to inquire before you start collecting.

It is up to your conscience to decide whether you want to take plants from their natural habitat. Certainly you should avoid thoughtlessly plundering plants. However, collecting small numbers of specimens is within the realm of conservation; in fact, it is one way of ensuring that these plants live on because most likely you will propagate the plants at home. Because vast areas of timberland may be bulldozed for building, the trees bromeliads live in may be destroyed. By collecting the plants you can save their lives.

Two: Bromeliads at Home

Bromeliads thrive at windows of every exposure, and when they bloom, you can move them about for table decoration or corner accent. I had a *Neoregelia carolinae* in my living room that was colorful for three months. The way you use your plants largely determines where to place them. I grow many of my bromeliads with my orchids, and others are on kitchen and bedroom windowsills in 3-foot-long plastic boxes. I put bricks in the boxes and set the pots on the bricks. This way it is easy to water, the drain-off is absorbed by the bricks, and the remaining water creates humidity.

INTERIOR DECORATION

Bromeliads make excellent home decoration because few are large or cumbersome; most Aechmeas, Billbergias, Guzmanias, Neoregelias, and Vrieseas grow to only 36 inches (91.4 centimeters) across and thus are never too large or ungainly in a room.

Some large Hohenbergias and Streptocalyxes can be used as vertical accents in a room or as room dividers or screens to partition an area. Generally bromeliads excel as accents at windows or on tables or desks. A group of bromeliads at a window never looks like an overgrown jungle; Guzmanias and Vrieseas are prime examples of fine plants that do not need much space to look handsome.

For desk accent or table beauty it is difficult to beat the smaller bromeliads; I think they are even prettier than the popular African violets. And in kitchens and bathrooms (favorite places for plants these days), bromeliads are right at home, providing a tropical character. Remember, bromeliads can live without very good light for many months.

You can use bromeliads for vertical accent anywhere in the home. Simply tack them to a board or to metal or wood trellises for a different and unique decoration. You can even pin smaller bromeliads, such as many Tillandsias, to shower curtains, where they will survive for months on the humidity in the air. Bromeliad leaves and roots can absorb moisture from the air and still survive.

CONTAINERS

There is an array of containers for plants in endless colors, shapes, and materials. I have found that the terra-cotta pot is best for bromeliads because moisture evaporates readily from the clay walls of the pots. The earth color blends into most situations, and the pots are not expensive. Plastic is lightweight, so tall bromeliads in plastic containers often topple over. And plastic holds moisture longer than terra-cotta (clay), which is not good for epiphytic plants.

Aechmea* growing in a bean pot makes a lovely house decoration. Bromeliads can go without direct light for weeks. *(Photograph by the author)

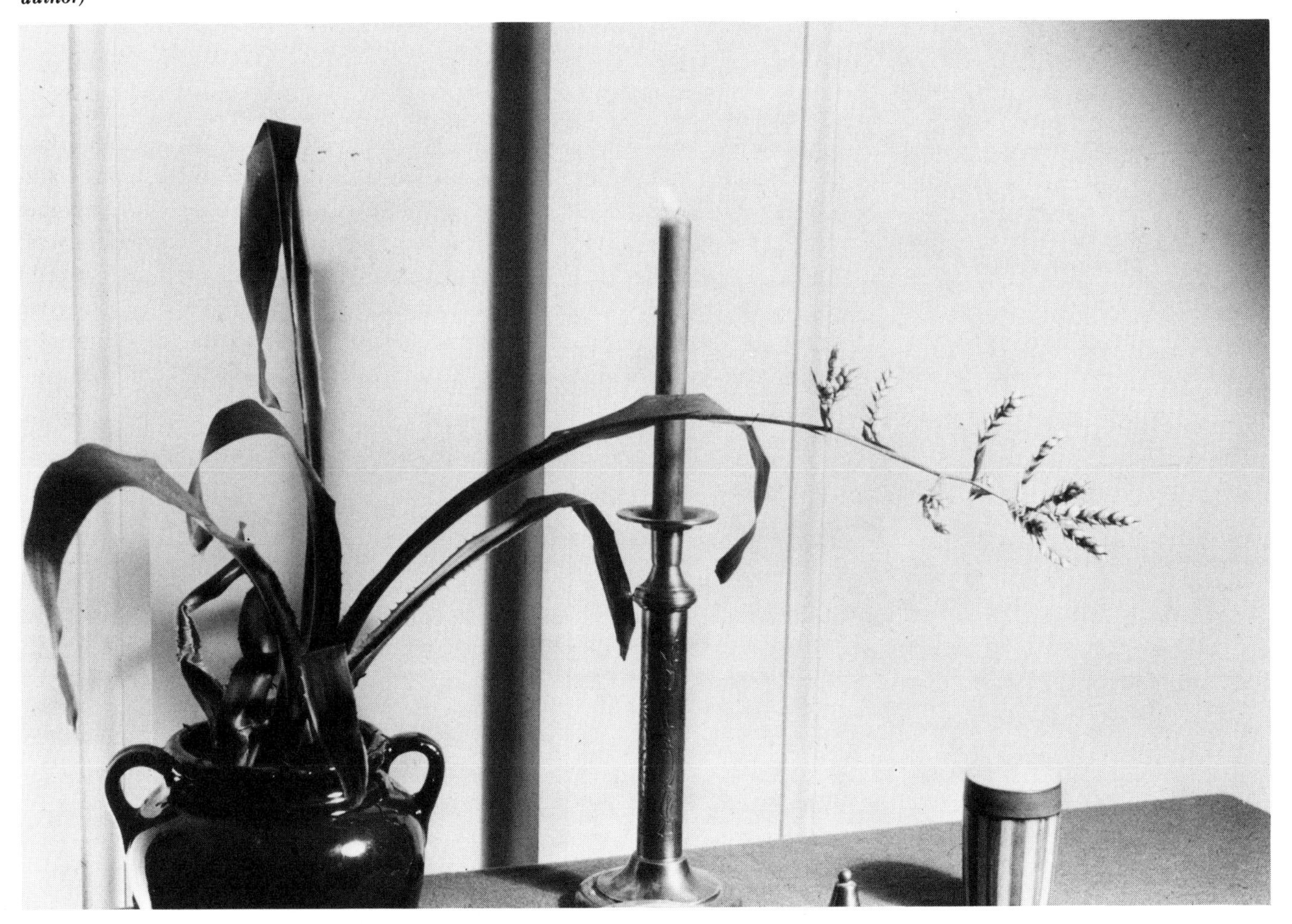

Pottery in different sizes and shapes can be used where a dramatic statement is needed. The plant at the left is a *Hohenbergia;* at right, a *Guzmania. (Photograph courtesy Architectural Pottery)*

Be wary of glazed pottery, pretty though it may be. The protective glaze prevents moisture from evaporating through the walls, which can create a soggy medium that is detrimental to bromeliads because they are air plants. Also avoid pretty pots that have no drainage holes; again, moisture will accumulate in the medium. If you have a decorative pot or jardiniere, use it only to hold a terra-cotta pot.

Wire baskets and wood containers are good for some bromeliads. Wire baskets are attractive, but unfortunately wood is not appealing in most homes, often clashing with rather than complementing the decor. However, slatted wooden containers can be handsome, and bromeliads grow very well in them because the open walls permit air to enter the growing medium, preventing it from becoming soggy. Slatted baskets are sold through mail-order suppliers. Potting in wire baskets and slatted wood containers takes some time because the inner shell of the container must be filled with osmunda to contain the potting medium.

Clay pots are the most suitable for bromeliads; they come in many sizes and water evaporates slowly from walls. ***(Photograph by the author)***

Small wooden containers found at nurseries, are fine for bromeliads, although wood seems to work better as a decorative accent outdoors. *(Photograph by the author)*

Plastic trays, available in various sizes, are also a great convenience, and sheet-metal companies make waterproof metal trays that can be fitted to windowsill space. Use a 2-inch-deep pan (5 centimeters) with rolled edges to avoid being cut. Fill the box with crushed stone to within ½ inch (1.3 centimeters) of the top; set the pots on top of the stone. Small bromeliads especially can be effectively grown this way. Water plants as you would if they were in plastic boxes.

For larger specimens, use 4-inch-deep (10 centimeters) metal trays on the floor under windows. Fill each tray to within ½ inch (1.3 centimeters) of

An arching *Acanthostachys* provides a graceful accent in a living room corner. *(Photograph by the author)*

the top with crushed stone and arrange the pots on the stone. Use a border of bricks to hide the edges of the trays. You can also use discarded radiator pans.

Bromeliads do not resent temperature and humidity variations; in fact, they readily adjust to such changes. Thus you can place each pot in a terra-cotta saucer and move plants about to suit your decorative needs. Cats and dogs are attracted to the succulence of most bromeliad foliage and delight in chewing the leaves, so keep your plants away from pets. Bromeliads can also be used decoratively in driftwood arrangements and as hanging plants on pieces of bark.

WINDOW ARRANGEMENTS

Do not crowd plants at windows or the garden will look unsightly. Use only a few handsome specimens, say, three or four to a window ledge. Protect

Aechmeas, Billbergias, Vrieseas, and Cryptanthuses (front) grow well together in a galvanized planter box filled with gravel chips and lined with brick. ***(Photograph by the author)***

wooden windowsills with cork mats, and elevate pots on pieces of tile or other waterproof material placed over the mats. Always insert a saucer under each pot to catch excess water.

If you have a standard-sized window (30 by 60 inches/76.2 by 152.4 centimeters), you can install glass or wood shelves across the window. You can buy packaged glass shelves at hardware stores or make your own shelves. Another possible window installation is shelves on pressure or suspension poles. These poles are held to ceilings and walls by compression, so no fixing hardware is necessary. Such poles and shelves can display six to eight containers in front of a window. The poles are sold at hardware stores. Similarly, you can display bromeliads on vertical wire or wrought iron vertical plant stands. These units have trays attached and are sold at florists' or gift shops.

Suspending plants at windows is still another way to color a room. Hanging containers are sold at garden suppliers and come with chain and ceiling hardware. Use four or five hanging containers to a window, and place the containers at different levels—high, low, and between—for a staggered design.

Several kinds of bromeliads in a galvanized, gravel-filled box make a spectacular indoor arrangement. ***(Photograph by the author)***

On driftwood bromeliads excel in dramatic statement and grow well too. Plant at left is ***Guzmania vittata;*** **right,** ***Neoregelia.***

Another driftwood display uses a *Tillandsia ionantha* at left and an *Aechmea* in center. Plants are wrapped in osmunda and wired in place. *(Photograph by the author)*

Almost any window in the home can be used for bromeliads, whether the exposure is east, west, south, or north. Select plants according to their needs. For example, Ananas and Billbergias need more light (east or south) than, say, Neoregelias or Nidulariums, which do fine at a north or west exposure.

MOUNTED BROMELIADS

Because so many bromeliads—Tillandsias and Guzmanias, for example—grow well and look handsome on pieces of tree bark, tree-fern fiber boards, or cork slab, you should know how to mount plants on these materials. Select a tree branch or small limb that has natural crevices or pockets. Fill the voids with osmunda (tree-fern root) and wire the osmunda in place, or use string. Place the bromeliad roots against the bed of osmunda and then wrap them in the osmunda. Now wire the bromeliad to the bark. Mist the plant and bark, and place the plant in bright light.

A bromeliad "tree" is a very decorative indoor accent. Use a small and

Small *Cryptanthus* bromeliads grow lavishly in a terrarium without a cover. *(Photograph by the author)*

A bromeliad "tree" grows on a piece of wood anchored into a dish with concrete. Plants are Tillandsias. *(Photograph by the author)*

Suspension poles are held at ceiling and floor with spring compressors and need no hardware attachments. Slatted shelves at various levels hold plants. ***(Photograph by the author)***

interesting tree branch as the "host" for the plants and an appropriate container to hold the tree. Center the branch and temporarily wire it in place. Then anchor it permanently with mortar (three parts sand, one part cement, and enough water as needed). Let the mortar harden for a few days. Then select the plants you want (Tillandsias and Guzmanias work best). Arrange the plants in a graceful curve on the tree branch; use plants of different sizes to provide interest. Vary color and leaf texture to create a handsome design. Affix plants to wood with osmunda. Select an interesting piece of wood that has natural pockets. Wire chunks of osmunda (tree-fern fiber roots) over the pockets, and then wrap more osmunda around the root crown; securely wire the plant in place or use string. In about six months the plant will send roots into the wood, at which time you can cut away the wire or string.

You can also make hanging arrangements with pieces of bark that have suitable pockets. Suspend the bark or tree cork from the ceiling with wire. These make handsome hanging decorations, and small Tillandsias and Guzmanias are effective used on wood.

Mounted bromeliads on wood or other materials (cork bark, tree-fern slabs) must be watered at the sink, so use chunks you can easily handle. Once or twice a week take the arrangement to the sink and water it thoroughly.

A window arrangement can include bromeliads (right and left) displayed on wooden trays attached to suspension poles. ***(Photograph by the author)***

(Of course, if your mounted plants are in an area with a tile or concrete floor, water can be applied directly to the plants.)

1. Wet osmunda.

2. Wire or tie with string to tree branch.

3. Place plant and wrap in osmunda. Wire or tie in place with string.

4. Water plant and bark.

Mounted bromeliads

ARTIFICIAL LIGHT

For people who have limited natural light in their apartments or houses, artificial plant lights are "miracles" because they enable beautiful plants to flourish in otherwise dim interiors. The more light most bromeliads get, the better the leaf color, and because so many of them have exquisitely colored foliage, bromeliads make excellent indoor artificial light subjects. Almost any small or medium-sized species can be grown successfully under lights.

Suppliers sell plant light carts or table models that can accommodate from three to fifteen plants. You can also make your own light setups.

If you do not want an elaborate setup with carts or stands, you can use the plant-growth lamp, shaped like a standard reading lamp, that fits into standard sockets. These inexpensive lamps are sold at suppliers.

For accent and maintenance lighting, use floodlamps, with the light directed at the plant. Appropriate wall or ceiling fixtures are necessary. Do not confuse accent floodlighting for drama with fluorescent light for growing.

A striking custom-made setup enables bromeliads to flourish under artificial light. ***(Photograph courtesy USDA)***

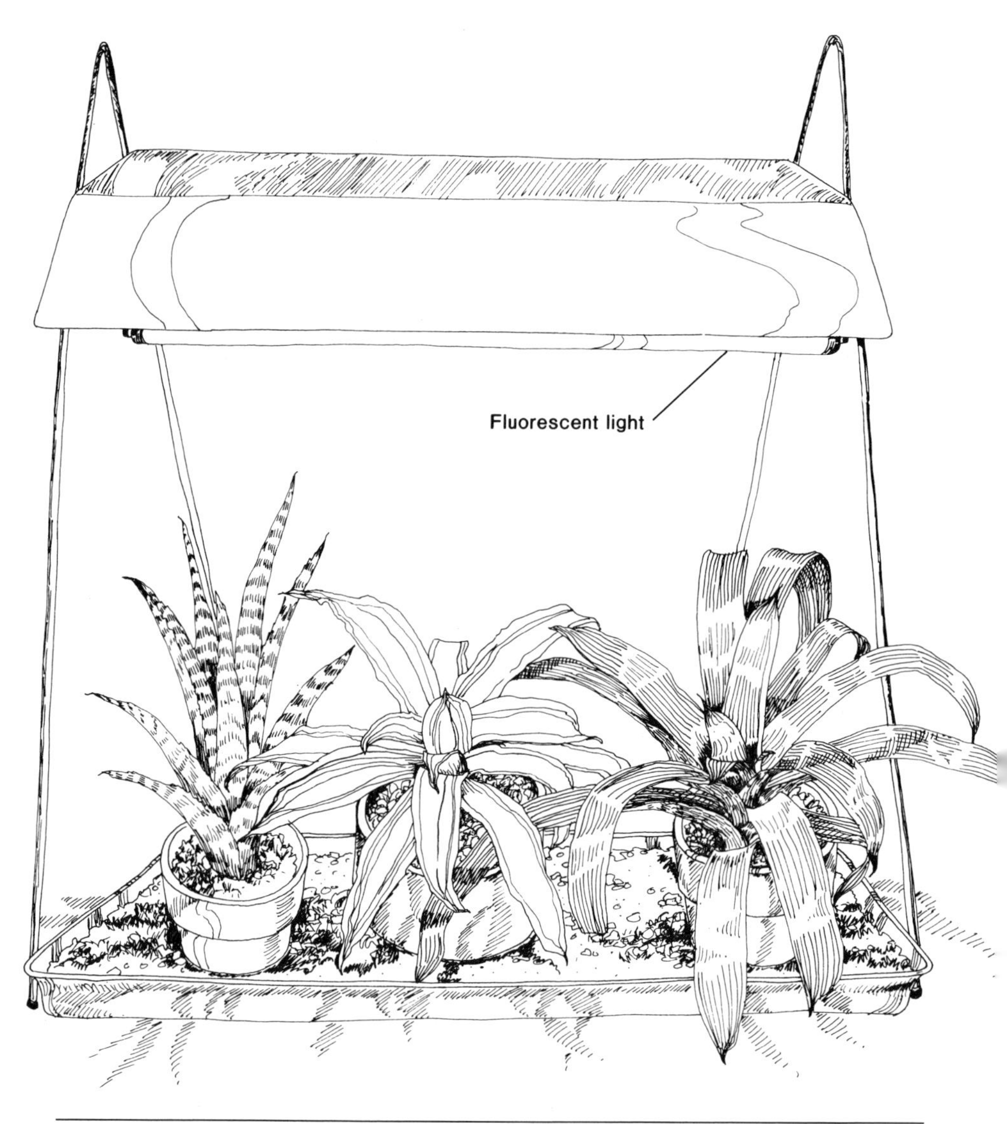

Bromeliads under lights

HOW PLANTS USE LIGHT

The visible spectrum of a rainbow has colors ranging from red to violet. Research indicates that plants require blue, red, and far-red rays to produce normal growth. Blue enables plants to manufacture carbohydrates; red controls assimilation of light and also affects the plants' response to the relative length of light and darkness. Far red works in conjunction with red in several ways: it controls seed germination, stem length, and leaf size by nullifying or reversing the action of red rays.

Plants grow best when they receive sufficient levels of blue and red light, which are in fluorescent lamps, and infrared, which is in incandescent lamps.

HOW MUCH LIGHT?

The intensity of fluorescent light given to a bromeliad must be tempered by common sense. Here are two helpful hints for better growing: For germinating seeds and growing seedlings, use 10 lamp watts per square foot of growing area. For high-light species like Hechtias and Dyckias, 20 lamp watts per square foot are beneficial.

If you use incandescent light in shelf gardens to furnish the vital far-red rays (infrared) that are lacking in most conventional fluorescent lamps, try the 4:1 ratio, which has worked well for me. For example, if you have 200 watts of fluorescent light, add 50 watts of incandescent light—five 10-watt bulbs. (Do not confuse these low levels of incandescent light with accent lighting, which is different.) Once or twice I tried increasing the incandescent ratio, but this only created heat that burned the plants. The 4:1 ratio is the norm but by no means the only possibility; trial-and-error is part of the adventure of growing plants under lights.

If the light intensity in shelf gardens proves too strong for some bromeliads, simply move them away from the light. Set the plants at the end zones of the lamps, where the light is less intense, or raise the adjustable reflector canopy. On the other hand, if light is not strong enough for some species, move them closer to the light. Put the potted plant on an inverted pot, or use a lattice support.

There is no set rule for how far a plant should be from fluorescent light, only general suggestions. Observe your plants; they tell you when they are getting too much light (the leaves are pale green) or when they are not getting enough (the leaves are limp).

FLUORESCENT LAMPS

Fluorescent lamps come in a dizzying array of shapes, sizes, voltages, wattages, and temperatures. Manufacturers use various names for their lamps: cool white, daylight, warm white, natural white, soft white, and so on. However, these names can be misleading because a natural white lamp does not duplicate the sun's light, a daylight lamp does not actually duplicate daylight, and there is no difference to the touch between a cool white and a warm white lamp.

Cool white lamps are closest in providing the kind of light—red and blue—necessary for plant growth. Daylight lamps are high in blue but low in red, and warm white and natural white, although high in red, are deficient in blue wavelengths. Fluorescent lamps come in 20, 40, or 72 watts.

In addition to these standard fluorescent lamps, several companies manufacture lamps designed solely for aiding plant growth. Among such lamps are Gro-Lux by Sylvania Lighting Company, Plant-Gro by Westinghouse Electric Company, and Vitima by Durolite Electric Company. Supplemental incandescent light is probably not needed with these lamps because plant-growth lamps have both red and blue quotients of light.

Besides the familiar standard lamp, there are newer lamps with high output. They may be grooved (Power Groove from General Electric Company), twisted (Powertwist from Durolite Electric Company), or they may show no difference in shape but be designated as high output (HO) or very high output (VHO).

Another development in fluorescent lighting is the square panel manufactured by General Electric Company. These cool white lamps give the same light as tubular shapes, but they have certain advantages. Attractive units can be created with them because the lighting mechanism is concealed, and vertical as well as horizontal lighting is possible. The panel lamp is 12 inches (30.5 centimeters) square and only 1½ inches (3.8 centimeters) thick. It fits into recessed, surface-mounted, or suspended units and comes in Panel Deluxe, Panel Deluxe Cool, and Cool White in 55 or 80 watts. The Panel Deluxe Cool type brings out vivid color hues, the closest match to natural daylight. Special electrical ballasts are required, and these lamps should be installed by an electrician. Plant-to-lamp distance should be 18 to 20 inches (45.7 to 50.8 centimeters).

INCANDESCENT LAMPS

Some authorities say that the major disadvantage of using incandescent lamps (reading bulbs) in a plant setup is that the heat they project can be too hot and drying for plants. Actually, an 8- or 15-watt lamp at a 10- or 12-inch distance does not produce enough heat to harm plants in shelf gardens. And in accent lighting, where 150 watts are used, the lamp source is placed 30 to 36 inches (76.2 to 91.4 centimeters) from plants—5 to 6 feet (1.5 to 1.8 meters) if mercury vapor lights are used. Again, there is no harm to plants from the heat.

Incandescent lamps are more expensive to use than fluorescent lamps; 70 percent of the power that goes into an incandescent lamp is wasted. However, when used as accent lighting for plants, incandescent lamps can keep a plant in a dim spot flourishing for years, whereas otherwise it would perish. So, if you have large ornamental greenery decorating your living room, incandescent floodlighting is well worth the extra cost.

Fixtures for incandescent lamps used for accent include bullet- or canopy-shaped reflectors that give a directional control of light to the plant and put it on display while providing beneficial light rays without an increase in heat. The bullet fixture is attractive and hides bare bulbs from view, and even one 150-watt lamp keeps a plant in a shaded area handsome for months.

Thus you can use (1) fluorescent lamps alone, (2) fluorescent and incandescent lamps together, (3) single-lamp bulbs, (4) standard flood lamps, and (5) the new mercury vapor lamps. No matter which lamps you use for your bromeliads, remember that none is a miracle worker by itself. Plants still need water, humidity, ventilation, and pest control. Some of the lamps may be better than others for plant growth, but plants will grow and prosper under any light if there is enough illumination and sufficient day length.

Three: Seasonal Culture and Care

Growing bromeliads in the home requires the use of specific cultural techniques. All varieties need light, a planting medium (bark or soil), containers, humidity, water, and occasional feeding. The plants also need some grooming and trimming. (Insect protection is covered in Chapter 4.)

PLANTING MEDIUMS

Many people grow bromeliads in soil, others use fir bark, and numerous gardeners grow plants in a combination of fir bark and soil. With air plants, the potting medium is merely something to hold the plant upright in the pot; bromeliads have shallow root systems and so derive few nutrients from any potting medium.

The fir bark you buy is the steamed, sterilized bark of redwood and cedar trees, although bark from other trees can also be used. Bark comes in small, medium, and large grades. Small-grade fir bark is used for plants to 20 inches tall (50.8 centimeters), medium-grade for plants to 40 inches (1 meter), and large-grade for bigger specimens.

Bromeliads can also be grown satisfactorily in osmunda, which is the root of a tree fern. If you want to grow bromeliads on tree-fern slabs or on wood, use osmunda to hold the roots. Some people use peat moss, but this can become a mess. Osmunda is sold in packages at orchid supply houses. It must first be soaked in water and then cut into chunks for easy handling. You can also pot bromeliads directly into osmunda.

If you use soil for bromeliads, select a soil that is neither heavy nor light

Bags of peat moss and bales of osmunda can be found in specialty plant shops. *(Photograph courtesy Matthew Barr)*

Osmunda fiber, packed into a kitchen gadget, creates a fine home for bromeliads. *(Photograph courtesy Matthew Barr)*

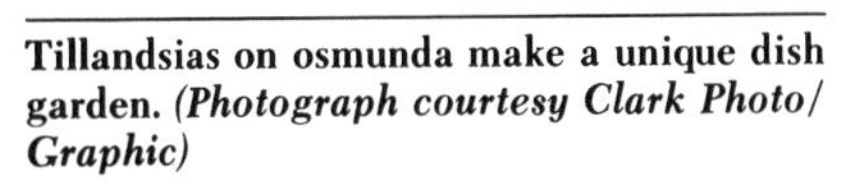

Tillandsias on osmunda make a unique dish garden. *(Photograph courtesy Clark Photo/Graphic)*

A tree branch sculpture combines osmunda and Tillandsias with natural fiber weavings into an art form. *(Photograph courtesy Clark Photo/Graphic)*

in weight. The soil should be porous so that air and water can move through it easily. You can use standard packaged potting soil, or make your own mix from one-third packaged soil, one-third humus (decayed organic matter), and one-third sand. This combination is medium in weight and drains readily.

Store bags of bark or soil properly. I once stored a 3-cubic-yard bag of bark in a somewhat sunny place; several weeks later it had mildewed because of heat and condensation within the sack. Store bark or soil in a dry, shady, and cool place.

WATER AND MISTING

If you can drink the water from your tap, plants can use it too. However, if you live in a region where they put excessive amounts of chlorine into the water supply, let the water stand overnight in a bucket to dispel some of the vapors; too much chlorine can sometimes be detrimental to bromeliads. Use tepid water when possible because icy cold water can shock plants.

When you water a bromeliad, really water it: soak it, let it drain, and then water again. Sparse watering causes air pockets in the medium, and then plant roots have to search for moisture, which weakens the plant. Water thoroughly until water drains from the bottom of the container.

If possible, water in the morning so that by evening everything is dry, precluding any mildew or fungus developing from too much moisture and no light.

Generally, water medium-sized plants—in 5- or 6-inch (12.7- or 15.2-centimeter) containers—three to four times a week in spring and summer, two to three times a week the rest of the year. Larger plants can go longer between waterings, and those in very small pots, such as *Tillandsia ionantha* and other tiny bromeliads, might need water every day in very warm weather.

As a general rule, water less during cloudy days and more on sunny days. Water less when plants are at rest, usually in the winter, than when they are growing in the spring and summer. (For more information about watering see end of chapter.)

Misting bromeliads helps increase humidity, even for a short time, and because bromeliads absorb water through their leaves, it helps plants grow. Misting also discourages insects, especially red spider mites, which proliferate in dry conditions.

To mist plants, use empty window-cleaner spray bottles that have been carefully rinsed out, or buy plastic misters especially made for plants. In hot weather, mist two or three times a day to keep humidity high. In fall and winter, mist only occasionally.

FEEDING

Some people feed bromeliads, but others do not. I prefer to take the middle-of-the-road approach, feeding plants about once a month in warm weather and not at all when cool weather starts. Too much plant food can harm the plants.

Plant foods are available in liquid, granular, dust, and time-released fertilizers. Buy a general liquid or granular plant food such as 10-10-5 and use it sparingly. However, I do suggest that you use fish emulsion for your bromeliads about once a month. Fish emulsion comes in a bottle and must be mixed with water—follow the directions on the label. It is an excellent source of nutrient for plants, neither too weak nor too strong.

LIGHT

Direct sun is beneficial for bromeliads because it turns their leaves a brilliant color. However, even without sun (a few hours a day), plants will be handsome as long as they receive some light. I mentioned that bromeliads can survive in all exposures—sun, bright light, diffused light, and even shade (north exposures)—but ideally most bromeliads should be at east or west windows because direct sun at southern exposures might scorch foliage.

HUMIDITY, TEMPERATURE, AIR CIRCULATION

Humidity for indoor plants is not the problem many people think it is because plants do not need a rain forest atmosphere to survive. In fact, too much moisture in the air coupled with dark days can cause rot. An average humidity of, say, 20 to 30 percent is fine for most bromeliads. Here are four ways to maintain good humidity indoors:

1. Set potted bromeliads on beds of gravel in planter boxes. Keep the gravel moist. Add water to replenish evaporated water, but be sure the pots do not actually sit in the water.
2. Grow many plants together (in groups) so they can generate their own humidity. As plants transpire (give off moisture) through their leaves, the condensation adds moisture to the air.
3. Spray plants with tepid water.
4. Install a small space humidifier.

If you are comfortable in your home, bromeliads will be too. An average home temperature of 70°F to 80°F (21.1°C to 26.7°C) by day is fine for most plants. Most bromeliads can survive during very hot days or very cold nights, but do not subject plants to a continually cold location or they will not grow well.

Air circulation is vital for most bromeliads. Whenever possible, keep a window slightly open near—not in—the growing area to provide a gentle flow of air at all times. In extremely cold weather, when you cannot open windows, use a small electric fan operating at low speed to help keep air moving.

GROOMING AND TRIMMING

All plants need some grooming to keep them attractive, and bromeliads are no exception. Leaves at the base of plants naturally die off and should be removed before they rot and cause fungus diseases. To remove a faded leaf from the base of a bromeliad, pinch it with your fingers and pull it loose and discard it. Leaves that have turned brown at the tips have usually been bruised—perhaps the plant was pushed against a window. Trim the leaves with small, sterile scissors. Keep bromeliad leaves clean and shiny—dust and soot settle on leaves and clog pores, so occasionally wipe foliage with a damp cloth.

Bromeliads should grow erect. If a plant begins to topple, firm the potting medium around the collar of the plant. You should stake tall tubular species to eliminate the possibility of plants toppling over. To stake a plant, insert a small stick at one side and plunge it into the soil, then tie the plant at its center with string to the stick. Do not use wire ties or you may cut the plant.

Remove faded flowers and bracts after plants bloom by cutting them off or by pinching them off with your fingers.

POTTING AND REPOTTING

Bromeliads can be potted in fir bark, osmunda, soil, or a combination of soil and bark. Most bromeliads do well in bark. To pot a bromeliad, select a clean terra-cotta container with drainage holes. If the pot is new, soak it overnight in water so it does not absorb moisture the plant needs. If it is an old clay pot, scrub it with hot water.

Bark: Use medium-grade fir bark. Some gardeners wet the bark overnight, but this can be messy. Dry bark works fine. Fill one-third to one-half of the clean pot with broken pieces of pots (shards). Set the plant into the shards and fill in and around it with fresh bark, occasionally pressing down the bark with a blunt-nosed stick or a piece of wood. Always work from the sides of the pot to the center until you have filled up the pot to within ½ inch (1.3 centimeters) of its rim. Most bromeliads need tight potting, with the bark firmly in place. If necessary, stake the plant with wood sticks and string the sticks to the plant. Label the plant.

Osmunda: Osmunda takes more time than bark to pot with because it is hard and dry. You must soak it overnight in water for easier handling. Fill the new pot with 2 inches (5 centimeters) of shards. Then cut the osmunda into small chunks, about 2 inches (5 centimeters) square. Set the plant into the pot, and fill in and around the plant with osmunda. With a potting stick, push the material down firmly toward the center of the pot. Trim away excess material with small scissors.

Soil and bark: Some bromeliads, such as Cryptanthuses, do best in a mixture of half leafmold and half bark. (You can use packaged soil for the leafmold.) The leaves of bromeliads are hard and durable; the weakest part of the plant is the core. If bromeliads are potted too tightly, water cannot drain away from the leaf cup or vase; if the vase holds water too long and the water does not drain off naturally into the compost, rot develops at the base. Gravel chips on top of the potting medium will help prevent this condition.

Repotting is removing a plant from an old pot and planting it in a fresh medium. Never pull a plant out of its container. First, rap the bottom of the

1. Put drainage shards in bottom of pot.

2. Fill halfway with fir bark or soil.

3. Center the plant. Fill in with soil, bark, or combination of both. Tamp down.

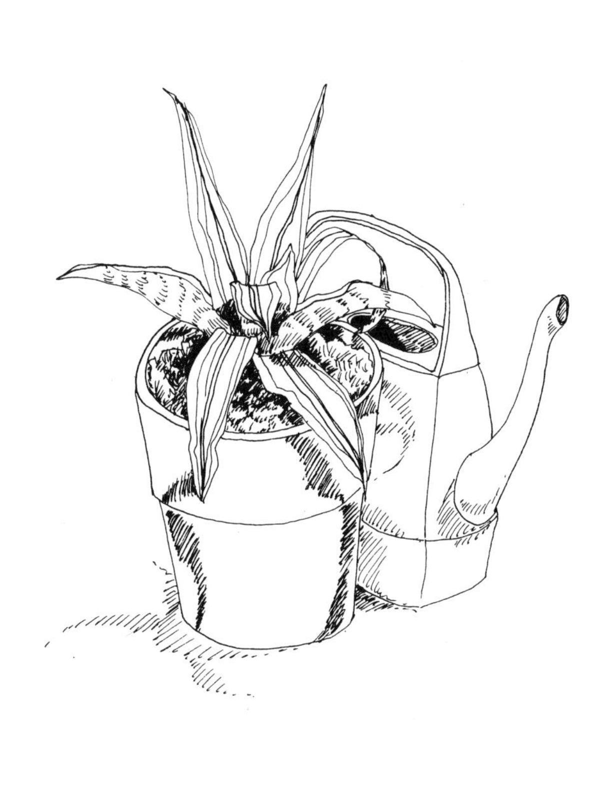

4. Water lightly and place in bright light.

Potting with bark or soil

container on a table or on the floor to loosen the plant's rootball. Then grasp the bromeliad at its base or collar and gently tug it loose. Wiggle it back and forth until it comes out of the pot without any pulling or force. Shake off the old potting material or crumble it off with your fingers. Now pot as described.

After potting or repotting, place plants in a temperature of 65°F to 70°F (18.3°C to 21.1°C) and out of direct sun. Wait about a week before watering, but spray daily with water. You can also mist the pot and the edge of the surface of the compost.

YEAR-ROUND SCHEDULE

SPRING

Water and light: Keep plants somewhat dry at the roots, but keep the cups formed by the vaselike growth of the plants filled with water. If the winter has been unusually dark, do not abruptly plunge bromeliads into sunlight; bright light suits them best now. (In summer they can be in sun.)

At this time of the year be sure bromeliads have ample ventilation. Keep a window open somewhere in the growing area, or run a small fan at low speed to keep air buoyant. Average indoor humidity of 20 to 30 percent is fine now. Spring rains will add to the humidity if you keep your windows open slightly.

Feeding: Most bromeliads react unfavorably to feeding, so it is best to let plants grow on their own as long as they have a fresh potting mix. However, if you have neglected to repot older bromeliads, feed them with a 10-10-5 fertilizer once a month.

SUMMER

Water and light: Keep the potting medium somewhat moist and the vases filled with water. Put the plants in sun, to aid mature growth and to encourage flowering.

Temperature and humidity: Provide good ventilation, which is vital in the summer when it can get very hot indoors. Be sure windows are open and there is a good flow of air. Mist plants frequently to maintain good humidity, and wipe leaves occasionally with a damp cloth to keep them shiny.

Feeding: It is tempting to feed plants now to force them into growth, but do not overdo it.

FALL

Water and light: Keep the potting medium evenly moist—water twice a week—and the vases filled with water at all times. The more light you can give bromeliads now, the better they will be next year, so select bright places for plants.

Temperature and humidity: As artificial heat is turned on, be sure to supply adequate humidity for the bromeliads. This is the best time to put

Acanthostachys strobilacea

Aechmea species (unidentified)

Aechmea angustifolia

Aechmea brevicolis

Aechmea chantinii 'Burgundy'

Aechmea chantinii 'Silver Ghost'

Aechmea fasciata

Aechmea chantinii x Aechmea ramosa

Aechmea fulgens discolor

Aechmea 'Meteor'

Aechmea mooreana

Aechmea fulgens discolor (flower head opening)

Aechmea mertensii

Aechmea fasciata variegata

Aechmea nudicaulis variegata

Aechmea zebrina

Aechmea tillandsiodes lutea

Aechmea 'Rajah'

Aechmea 'Red Wing'

Aechmea 'Spring Beauty'

Ananus comosus

Billbergia pyramidalis concolor (closeup)

Billbergia elegans

Billbergia brasiliensis

Billbergia nutans *Billbergia pyramidalis concolor*

Billbergia lietzii *Bromelia serra variegata*

Billbergia vittata

Canistrum cyathiforme (closeup)

Canistrum lindenii

Canistrum cyathiforme

Cryptanthus 'Minibel'

Cryptanthus fosteriana 'Elaine'

Group of *Cryptanthuses* on driftwood

Cryptanthus 'It'

Fasicularia pitcairnifolia

Guzmania* 'Magnifica' *(lingulata x minor)

Guzmania lingulata major

Guzmania zahnii

Guzmania lingulata minor

Guzmania 'Symphonie'

Guzmania 'Orangeade'

Hohenbergia stellata

Neoregelia compacta

Neoregelia carolinae

Guzmania 'Minnie Exodus'

Neoregelia carolinae var. tricolor

Neoregelia carolinae var. tricolor (closeup)

Neoregelia carolinae 'Meyendorfii' *variegata* (closeup)

Neoregelia carolinae 'Meyendorfii' *variegata*

Neoregelia cruenta

Neoregelia 'Marmorata'

Neoregelia spectabilis

Neoregelia 'Purple Passion'

Neoregelia 'Red Knight'

Nidularium innocentii lindenii

Nidularium innocentii var. lineatum

Neoma **'San Diego'**

Orthophytum navioides

***Orthotanthus* 'What'**

Portea petropolitana extensa

Quesnelia marmorata (closeup)

Quesnelia marmorata

Tillandsia caulescens

Tillandsia circinnata

Tillandsia cyanea

Tillandsias on driftwood

Tillandsia species (unidentified)

Tillandsia fasiculata

Tillandsia ionantha

Vriesea barilletii

Vriesea reginae

Vriesea carinata aurea

Vriesea 'Rubin'

Vriesea petropolitana

Vriesea splendens 'Meyer's Favorite'

plants on pebble trays (keep the pebbles moist) and to use a small space humidifier if you have one.

Feeding: Do not feed plants at all; bromeliads react unfavorably to most plant foods. These are nature's air plants, so let them grow naturally.

Note: Get your plants ready for the winter by removing all dead leaves and flower bracts.

WINTER

Water and light: Keep plants in the best possible light—sun is preferable. If you cannot provide natural light, use artificial lamps (even a single lamp will supply light for the plants). Keep the growing medium moist at all times.

Temperature and humidity: Keep bromeliads in temperate conditions; never let temperatures go below 55°F (12.8°C) in the growing area, and mist plants frequently because rooms are likely to be very dry from artificial heating.

Feeding: Do not feed plants at all.

PROPAGATION

Propagation means producing more plants from the ones you have. Bromeliads bloom only once in their life cycle and then die, so knowing how to propagate plants from offshoots is an essential way of making that original bromeliad immortal. You can get new bromeliads from mature ones by potting the offshoots (kikis) that grow from their bases. Aechmeas, Billbergias, Neoregelias, Nidulariums, Quesnelias, Canistrums, and other bromeliads produce these offshoots. (Tillandsia offshoots should be left on the mother plant rather than cut off.) Some offshoots grow on stolons (runners) that grow out horizontally from the parent, others grow at the base of the plant, and still others grow slightly off-center from the plant. You can also get more plants from bromeliads by dividing their large clumps; many species, such as *Aechmea calyculata* and several Billbergias, bear so many offshoots that they form large clusters which can be divided.

Using offshoots is the most common and quickest way to propagate plants. However, some gardeners like to sow seed, but this takes time and patience because seeds may take as long as six years to mature.

OFFSHOOTS

When a young offshoot is one-third the size of its parent and has about five leaves, you should pot it separately to enhance its growth. Cut off the offshoot with a sterile knife, or simply twist it at the base until it loosens and comes off its parent. Pot the young plant in your favorite mixture and water it and put it in bright light.

Several Vrieseas produce only one offshoot, off-center from the parent plant. Let this offshoot grow into the main plant as the parent dies off. Then remove the parent plant and discard it; leave the new plant growing in the same pot.

1. Wait until the shoot is 3″ tall.

2. Sever with a sharp knife.

3. Pot in bark or soil.

4. Put in plastic bag for a few days.

Offshoot propagation

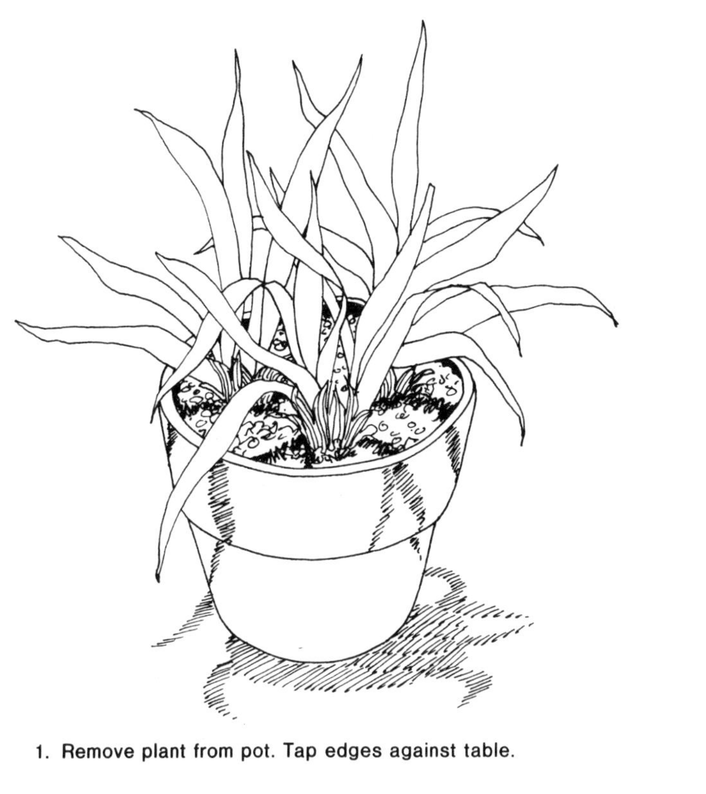

1. Remove plant from pot. Tap edges against table.

2. Turn plant on side and cut in half.

3. Separate into two or three small clumps.

4. Place each new plant in separate pot.

Division

DIVISION

If you have a cluster of Aechmeas or Billbergias that is too large for its present container, divide them. First rap the rim of the pot and remove the rootball. Then, with a knife, cut the rootball vertically, making two or three equal-sized clusters of plants. Repot each group.

SEEDS

Growing bromeliads from seed takes time, but it has its rewards when you can say, "I grew that myself." Pollinated flowers of bromeliads develop pods that contain seeds inside. Usually it takes two to three months for the seed of Aechmeas, Billbergias, Neoregelias, Nidulariums, and other bromeliads to ripen. Pick the pods when they are soft (ripe), and squeeze out the seeds onto pieces of paper. Move the seeds around until they dry, at which point they can be separated. The capsulelike pods of Tillandsias and Guzmanias often take as long as a year before their seeds mature—in this case you must wait until the pods burst and release the seed.

To grow seeds successfully, use a closed container. Any terrarium with a top or a plastic box will do. Start seed in vermiculite or any sterile starting medium; put a ½- to 1-inch bed (1.3 to 2.5 centimeters) of the medium in the bottom of the container. Carefully arrange the seed on top of the medium and then sprinkle the medium with water. As you sprinkle, be careful not to dislodge the seeds; just let the water trickle into the starting medium.

Close the container and put it in a warm (78°F/25.6°C) place until the seeds start to germinate (sprout fresh new growth). A few days after germination starts, move the container to a brighter place. If much moisture condenses within the container, remove the cover for a few hours during the day. When the seedlings are about 1 inch (2.5 centimeters) high, remove them with the end of a pencil. Place each seedling into a 1- to 2-inch (2.5- to 5-centimeter) pot of potting medium. Keep the medium just moist, never dry or soggy. When the plants are about 3 inches (7.6 centimeters) high, pot each one in a separate container of equal parts of soil and bark or any other suitable medium.

Tillandsias are the exception to standard seed-starting methods. Germinate seeds of these plants on osmunda or sphagnum moss wired to tree slabs or bricks. Do not cover the slabs or bricks or put them in a closed case because Tillandsia seed needs oxygen to germinate successfully. When seed germinates, pot as described for other seedlings.

You can also grow seed under artificial light. This is a fine way to start plants. Consult the earlier section on artificial light if you plan to propagate bromeliads this way.

FORCING FLOWERS

Some species of bromeliads may be reluctant to bloom indoors, even if light is good. If a mature plant has not blossomed, you can stimulate flowering by giving the plant ethylene gas. Use ripe apples, which emit this gas. Put the entire plant—pot and all—in a plastic bag with a ripe apple. Be sure to empty the cup of the plant because too much moisture while the plant is in a closed bag can harm the plant. Tie the bag and place it in bright light (but no sun) for about seven days. Then remove the bag and throw away the apple; in about a month the bromeliad should bloom.

1. Wait until pods are slightly soft. Break pod carefully to extract seeds.

2. Roll on newspaper to separate seeds and let dry.

3. Plant in a terrarium filled with ½″ vermiculite. Cover and keep in bright warm place.

4. When seedlings are 2″ high, remove and pot in 2″ pot.

Seed propagation

Bromeliads can be forced into bloom sealing the plant in a plastic bag with a ripe apple. ***(Photograph courtesy USDA)***

Four: Insects and Other Problems

To you, plants are a visual treat, but they can be a gastronomical treat to insects. Not many insects will bother bromeliads, but there may be occasional visitors. Some can be eliminated quickly, but others persist, no matter what precautions you take. Pest problems are minimal in the winter because insects proliferate in the spring and summer; the warmer the weather, the more broods of hatching insects there will be. But insects are no problem if you tackle them early. However, when left to multiply, insects cause havoc, so observation is nine-tenths of the battle.

By far the peskiest insects are mealybugs, which are white, cottony, and waxy. They can multiply overnight: a female mealybug is capable of giving birth to 600 babies in 48 hours! Red spiders occasionally attack bromeliads. You can prevent an invasion by not keeping plants moist in hot and dry weather. Aphids (plant lice) will bother your plants if you do not kill them early. Scale insects attach themselves to plant stems and suck plant juices, but they are not as persistent as mealybugs.

You should know where to look for insects because they attack various parts of plants. Most insects lurk in the leaf axils and on the undersides of leaves. Scrutinize these places carefully, with a magnifying glass if necessary. Scale almost exclusively cling to stems at the lower part of a plant.

Before the hot summer weather sets in, wash bromeliads at the sink with a strong spray of water. Gently massage the leaf axils with water, and then wipe the leaves so they are shiny and free of soot, dust, and any spores that might harbor enemies.

Another important procedure for insect prevention, especially in sum-

Aphids occasionally find their way to bromeliad foliage, but generally leaves are too tough and they migrate to other plants. ***(Photograph courtesy USDA)***

mer, is to soak pots in the sink in water up to their rims; if any insects are lingering in the soil, they will come to the surface after a few hours.

Most insects are recognizable on sight; red spiders weave telltale webs. Here is a general description of the pests and the damage they do.

Aphids: These small, pear-shaped, soft-bodied insects have beaks and four needlelike stylets. They are brown, red, green, or yellow. Plants lose vigor and become stunted; leaves curl or pucker.

Mealybugs: These cottony, waxlike insects have segmented bodies. They wilt leaves and stunt young growth.

Red spider mites: These are tiny, oval-shaped, yellow-green, red, or brown creatures. They cause foliage to become pale and stippled.

Scale: These are small insects with armor-type shells. They cause leaf and stem damage.

PREVENTIVES

If you look at your plants daily and catch insects early, damage will be minimal. If you see the insects on a few leaves, cut off the leaves. However, in most cases some more precaution is necessary, preferably natural, old-fashioned preventives or chemicals.

Mealybugs, common pests of most plants, only rarely attack bromeliads. ***(Photograph courtesy USDA)***

OLD-FASHIONED REMEDIES

1. **Laundry soap and water.** Mix half a bar of a laundry soap like Ivory with 5 quarts (4.7 liters) of water. Dip a cotton swab into the mixture and dab the insects. Or, at the sink, pour the soap-and-water mixture over the plant. Then rinse the plant with tepid water. Repeat at three-day intervals.
2. **Alcohol.** Put alcohol on a cotton swab and dab the swab directly on mealybugs and aphids to kill them. Repeat applications as necessary.
3. **Tobacco.** Steep old tobacco from cigarettes in water for several days to make a solution that will get rid of scale. Repeat every third day.
4. **Handpicking.** Handpick insects off with a toothpick.
5. **Wiping.** Wipe leaves with a damp cloth dipped in plain water to eliminate any insect eggs before they hatch.
6. **Water spray.** This may sound ineffective, but if you persist, it works on many insects. Use a strong spray of water at the sink.

INSECTICIDES

The use of poisons in the home can be hazardous. If you use chemicals rather than the old-fashioned remedies, know something about them to avoid confusion (there are so many on the market) and possibly killing your plants.

Insecticides are available in many forms. Water-soluble insecticides are sprayed on plants with special sprayers. Powders or dusts are not necessary in the home. Systemics—granulated insecticides applied directly to the soil—are very convenient to use: Spread the granules on the soil, then thoroughly water the plant. The insecticide is drawn up through the plant's roots into the sap stream, making the sap toxic. Thus, when sucking and chewing insects start dining on the plant, they are poisoned. Systemics protect plants from most, but not all, sucking and chewing pests for six to eight weeks, so generally you have to apply this type of insecticide only three or four times a year.

If you use chemicals, follow the directions to the letter or the chemicals will harm you as well as the plants. Usually repeated doses are necessary to fully eliminate insects. *Keep poisons out of the reach of children and pets.* A good, general chemical that does not have a cumulative effect is Malathion. Whichever insecticide you use, always follow these six rules:

1. Never use a chemical on a bone-dry plant.
2. Never spray plants in direct sun.
3. Use sprays at the proper distance marked on the package.
4. Try to douse insects if they are in sight.
5. Always use chemicals in well-ventilated areas; outdoors is the best.
6. Read and follow manufacturer's directions on applying chemicals and on protecting yourself while doing so.

Here is a summary of insecticides and the insects they kill. (Aerosol bombs are generally sold under different trade names as indoor plant sprays. They can harm leaves if you spray too closely and can irritate lungs. Do not use an outdoor spray for indoor plants.)

Trade or Brand Name	Insects	Remarks
Malathion	Aphids, mites, scale	Broad-spectrum insecticide; fairly nontoxic to human beings and animals
Diazinon	Aphids, mites, scale	Good, but more toxic than Malathion
Sevin	Most insects	Available in powder or dust forms
Isotox	Most but not all insects	Systemic; toxic but effective
Meta-Systox	Most but not all insects	Systemic; toxic but effective
Black Leaf 40	Aphids and sucking insects	Tobacco extract; relatively toxic
Pyrethrum	Aphids, flies, household pests	Botanical insecticide; generally safe
Rotenone	Aphids, flies, household pests	Used in combination with Pyrethrum; safe

PLANT DISEASES

If bromeliads are well cared for, they rarely develop a disease like fungus or botrytis. If they do and are treated early, bromeliads can be saved, but if left unchecked, they can die from plant diseases.

Plants show signs of disease by visible symptoms like spots, rot, and mildew. Many plant diseases cause similar external symptoms, so you must identify the specific disease to be sure you apply the correct remedy. Your state agricultural agent, listed in the Yellow Pages, is your best source of help. Too little or too much humidity can help cause disease, but diseases are mainly caused by bacteria and fungi. Bacteria can enter a plant's naturally minute wounds and small openings. Once inside, bacteria multiply and start to break down plant tissue. Animals, soil, insects, water, and dust carry bacteria that can attack plants. And if you have touched a diseased plant, you too can carry the disease to healthy plants. Soft roots, leaf spots, wilts, and rots are four diseases caused by bacteria.

Fungi, like bacteria, enter a plant through a wound or a natural opening or by forcing their entrance directly through plant stems or leaves. Spores are carried by wind, water, insects, people, and equipment. Fungi multiply more rapidly in shady, damp conditions than in hot, dry situations because moisture is essential in their reproduction. Fungi cause rusts, mildew, some leaf spots, and blights.

FUNGICIDES

Fungicides are chemicals that kill or inhibit the growth of bacteria and fungi. They come in ready-to-use dust form, in wettable powder, or in soluble forms that mix with water and are sprayed on. Here is a brief résumé of the many fungicides available:

Aineb: Used for many bacterial and fungus diseases.

Benomyl: A systemic used for certain bacterial and fungus diseases.

Captan: A fungicide that is generally safe and effective for the control of many diseases.

Ferbam: A very effective fungicide against rusts.

Karanthane: Highly effective for many types of powdery mildew.

Sulfur: An old and inexpensive fungicide that controls many diseases.

Five: Bromeliads Outdoors and in Greenhouses

The beauty of bromeliads does not have to be limited to the indoors. In Chicago I had bromeliads in the garden until October; they did much to create a tropical atmosphere and bring diversion to the garden scene. In northern California, bromeliads can stay outdoors most of the year because temperatures rarely fall below 40°F (4.4°C).

You can dress up a terrace or patio with bromeliads, and if you live in southern California or Florida or other year-round frost-free climate, you can use them in trees, in much the same way they grow in nature, providing a unique garden. You can also grow bromeliads in greenhouses or solariums for a handsome display.

GARDENS

For unusual garden decoration, strategically place potted bromeliads in flower beds and other color areas. Do *not* take them out of their pots and plant them in the ground. Leave bromeliads in their containers so you can bring them indoors when cold weather comes. Generally this is late September in most regions. A mass of Neoregelias at ground level provides a panorama of color that will soothe the eye and the soul. Plants are easy to care for outside, and if there is ample rain, they need little more than an occasional check for insects, which will be more prevalent outdoors than in. Wash down foliage frequently to discourage any pests; use the suitable preventives discussed in Chapter 4 if insects attack.

If you live in a climate where it is possible to grow bromeliads on low branches of trees, wire the plants to the trees. Lay a bed of osmunda—about

2 inches (5.1 centimeters)—on the tree branch, and wire it in place with standard galvanized fine-gauge wire, sold at hardware stores. Then set the bromeliads' roots into the osmunda and cover them with another bed of osmunda; wire this second bed in place. Eventually the roots will grow well into the osmunda and grasp the tree firmly; when that happens, the wire can be cut away. It takes about a year for a bromeliad to anchor itself firmly to the branch; the plants will thus eventually become permanent residents of the garden.

In all-year temperate climates, plants will need ample moisture—spray

A group of Neroegelias add an unusual but attractive note in a California backyard where temperatures rarely fall below 40°F (4.4°C). ***(Photograph by the author)***

Putting bromeliads outdoors during warm weather adds spot color to the garden. Guzmanias and Neroegelias in pots decorate this yard. ***(Photograph by the author)***

them with water daily. Otherwise, care is minimal. For a good effect use mature plants in groups. This creates a concentration of color rather than a spotty effect.

PATIOS AND TERRACES

Because bromeliads are naturally epiphytic and prefer arboreal positions, you can hang plants on almost any construction members of a patio or terrace: posts, columns, house walls, eaves, and so forth. Wire the plants to these members; bring plants indoors in the winter. You can also use potted bromeliads as spot decorations on the patio or terrace. Use the plants in groups to create a concentrated mass of color. Potted bromeliads also should be brought inside when cold weather starts.

Bromeliads provide a dramatic accent for a patio or terrace when grown on trellises. These decorative structures can create charm in a garden; you can make a portable hinged-screen trellis that can be set outdoors in warm weather and brought indoors in the fall and winter. To attach brome-

Lining a walk in a Chicago garden is a group of bromeliads in pots: *Vreisea splendens* is the accent. In October the plants are dug out of the ground and returned to the house. *(Photograph by the author)*

liads to trellises, follow the same procedure as for putting them on trees; that is, secure them in place with wire on beds of osmunda.

RETURNING PLANTS INDOORS

When cold weather comes, return outdoor bromeliads to their indoor spots. However, before you do so, inspect the plants for insects, to be sure you are

A beautiful *Aechmea fasciata* is part of a California garden scene. Plants stay outdoors most of the year. *(Photograph by the author)*

Guzmanias and Aechmeas enhance a patio, providing fine color. *(Photograph by the author)*

In planter boxes along a swimming pool, *Aechmea fasciata* and a fine *Guzmania* provide color. *(Photograph by the author)*

not bringing any pests indoors. Spray down plants with a heavy stream of water, and set pots in pails or sinks full of water to eliminate any critters.

The outdoor rains and cooling breezes do wonders for bromeliads, helping them survive indoors during the gray days of winter. It is work to move plants from indoors to out and then back again, but it is a way to create a novel effect outdoors and enjoy the plants all year.

The California pool landscape looking from the other direction, where more bromeliads brighten the picture. ***(Photograph by the author)***

This gardener summers his bromeliads on the porch steps—good for color, good for the plants. ***(Photograph by the author)***

A garden room/greenhouse boasts many fine, thriving bromeliads. Tillandsias are on a hanging raft at left. ***(Photograph by the author)***

GREENHOUSES

Many gardeners say that in a greenhouse there are ideal conditions for growing plants, and there is some truth to this. However, greenhouse growing of

An unusual greenhouse area is aglow with bromeliads, with dozens of Neroegelias at floor level. ***(Photograph courtesy Max Eckert)***

any plants, including bromeliads, requires somewhat different cultural concepts than growing plants in the home. In a greenhouse there are more light and more humidity, so plants will grow faster and more lushly. But greenhouse growing does have some drawbacks.

Under glass, plants can get very hot, and too much heat desiccates plants unless there is ample moisture, so keep your greenhouse plants well sprayed with water and wet down floors daily in warm weather to provide the extra humidity that is needed. Be sure the light is not too strong; summer sun can be intense, and I have had leaves scorch on very hot days. Shade the greenhouse in peak summer weeks with whitewash or plastic screening.

The author checking specimens in the greenhouse at the University of California in Berkeley, where bromeliads and orchids grow side by side. *(Photograph courtesy Joyce R. Wilson)*

In winter, heating a greenhouse is very costly; you can grow plants at 45°F to 50°F (7.2°C to 10°C) to conserve heat and cost because bromeliads can tolerate these temperatures, although they may not like it that much. However, at such times, when temperatures are cool and light sparse, remember to cut down watering and to keep the plants barely moist. Always keep the vase of the plant filled with water. Feeding is not necessary during cold months.

A greenhouse creates high humidity; in the winter keep checking for fungus diseases, which can start in such conditions. Remember to let air enter the glass garden because a good circulation of air helps prevent insects and disease from starting.

Of course the greenhouse should always be scrupulously clean—no debris, dead leaves, or flowers on the floors or benches. Keep the greenhouse spotless or disease and insects will start. Because a greenhouse is usually crowded, insects and disease can move from one plant to another quickly.

Part 2/A WORLD OF BROMELIADS

Six: Over Two Hundred Bromeliads for You

In the following pages are more than 200 different kinds of bromeliads to grow—there are small ones, medium-sized plants, and large specimens. The plants are arranged alphabetically within each genus, with general notes and specific cultural directions followed by a description of the more popular species.

While sizes of plants are given either in leaf length or rosette, these are not to be taken as absolute sizes. Growth depends upon many factors and many Bromeliads are variable in their size. The dimensions given are those of the plants I grew under my conditions.

ABROMEITELLA (a-brom-eet-EL-a)

A genus of only two plants, these bromeliads resemble Dyckias but are smaller, with rosettes up to about 2 inches (5.1 centimeters) in width. The pretty leaves are stiff and gray-green and form clumps of plants, making them most unusual in appearance. As houseplants they do well in decorative containers and offer a different look. Plants are from Argentina and Bolivia, where they grow in dense clumps on rocks in hot, dry areas.

Grow these oddities in a well-drained medium, such as equal parts of soil and bark. Be sure drainage is perfect because they will not tolerate stagnant conditions at their roots.

***A. brevifolia** (brev-ee-fol-ee-a)* has small 4-inch (10.2-centimeter) rosettes of pointed, stiff, gray-green leaves forming a cluster; greenish-white flowers are borne from the center of each rosette.

CULTURE

Water: Keep somewhat on the dry side; overwatering can harm the plant.

Potting medium: Soil or a bark-and-soil mixture; use small pots.

Feeding: Twice a year with fish emulsion.

Light: Likes sun but will tolerate bright light.

Temperature: Likes hot temperatures, but will tolerate cool nights.

Propagation: By offshoots.

Remarks: This is a graceful, unusual table plant that is good if you want something different.

Abromeitella brevifolia

A. chlorantha

A. chlorantha *(klor-an-tha)* has tiny rosettes, about 2 inches (5.1 centimeters) across, of scaly and fleshy 1-inch (2.5-centimeter) silver-green leaves. Forms tiny mats. Unusual.

ACANTHOSTACHYS (a-can-tho-STEAK-is)

Until recently this group of bromeliads had only one species available; lately, however, a few other species have been introduced. The plants have good decorative value because they are unusual, with graceful and arching thin, spiny leaves. These plants are ideal when grown in hanging baskets.

Found in southern Brazil, Paraguay, and Argentina, Acanthostachyses grow mainly on rocks and need little water and light to prosper, making them ideal for the forgetful gardener. Acanthostachyses certainly cannot match the beauty of Aechmeas or Billbergias, but they are fine small plants for table and desk accent.

A. strobilacea *(stro-beel-a-cee-a)* has reddish-brown, pendant, whiplike leaves to 26 inches (66 centimeters). The small, orange, pineapple-type inflorescence is surrounded by stiff bracts. A graceful and handsome plant.

CULTURE

Water: Keep somewhat dry; never overwater. Let soil dry out between waterings.

Potting medium: Soil or bark and soil.

Feeding: Occasionally during the summer.

Light: Sun or bright light.

Temperature: Grows well at average to warm temperatures. 60°F to 80°F (15.6°C to 26.7°C).

Propagation: By offshoots.

Remarks: This unusual plant with graceful foliage does well indoors.

AECHMEA (EEK-mee-a)

With almost 200 species, Aechmeas are the best known of all bromeliads. They are readily available and have beautifully colored foliage. Also, these plants produce flower crowns that last for months; recent varieties have put these bromeliads at the front of the pack, with one plant more handsome than another.

Plants range in size from 24 inches (61 centimeters) to about 5 feet (1.5 meters) in diameter, and their shape is a tubular rosette—almost a funnel type of growth. The leaves, edged with tiny spines, may be plain green but more often are green banded with handsome silver, maroon, rose, or purplish brown. Some species have leaves blotched or striped with color rather than banded, and a few have leaves with a combination of all three markings. The inflorescence may be branched, cylindrical, or pendant, as in *A. racinae*. The flowers themselves are usually small; it is the bracts that produce the vivid colors and last for weeks. Most Aechmeas bear berrylike, brilliantly colored fruit after they flower. As a rule, these plants are very free with offshoots, frequently throwing three or four "kikis" from a mature specimen.

The plants are native to central Mexico south to Argentina and most grow on trees or rocks while some species inhabit the forest floor. It is not uncommon to find them at 6,000 feet (1,829 meters) in bright sunlight or at sea level in dappled shade. From my experience, the tree dwellers outnumber the terrestrial forms.

Under cultivation, Aechmeas prefer a bright sunny place. I pot my plants in equal parts of fir bark and soil; this medium offers excellent drainage. I keep the growing medium evenly moist all year, and I do not worry about watering the plants regularly. If they go a week without water, they will survive. However, it is vital that there always be water in the vase of the plant because this is their lifeline; if the vase is left dry, plants suffer.

CULTURE

Water: Keep cups of plants filled with tepid water at all times. Keep the potting medium evenly moist, never soggy and never dry. Water plants in 6- to 10-inch (15.2- to 25.4-centimeter) pots two to three times a week in warm weather, once or twice a week the rest of the year.

Potting medium: Use a potting mix of equal parts of medium-grade fir bark and soil. Put a few gravel chips on top of the mix.

Feeding: Use a mild fertilizer such as 10-10-5 every third watering. Do not overfeed.

Light: Moderate to bright; some sun is fine.

Temperature: Will tolerate temperatures to 45°F (7.8°C). Good growing temperature is 75°F (23.9°C) by day, 65°F (18.3°C) at night.

Propagation: Offshoots root readily; remove them from the base of a plant when they are 2 to 4 inches (5.1 to 10.2 centimeters) tall.

Remarks: Excellent room plants for spot decoration on tables and desks. Occasionally wash leaves with a damp cloth.

The sunny location is best, but some of my Aechmeas do well in a north light too, although their foliage is not as brilliantly colored. Most bloom without too much pampering if light and moisture are good. The smooth-leaved Aechmeas can be wiped with a damp cloth occasionally, but those with scaly leaves (and a handsome powdery coat) should not be handled because they get marred when touched. Most plants can be misted with tepid water to keep humidity high. It is essential that Aechmeas be in a location with excellent air circulation or they will not do well.

During all my years of growing Aechmeas I have never seen any insects on the plants, and they seem to be impervious to pests and disease as well, making them stellar houseplants. Their one disadvantage is that tall species such as *A. luddemanniana* and *A. chantinii* when filled with water become so topheavy that even clay pots can fall over. Also, the base of the plant may become weak and bend; I tie my Aechmeas that do this to small sticks, using string.

A. angustifolia *(an-goos-tee-fol-ee-a)* grows to about 24 inches (61 centimeters) and has narrow leathery leaves. White berries appear after the spring blossoming. In a few weeks they turn brilliant blue and remain on the plant for two months.

A. bracteata *(brack-tee-a-ta)* is a very robust plant, becoming bottle-shaped, 3 to 5 feet high (91.4 centimeters to 1.5 meters); glossy, apple-green leaves

Aechmea angustifolia

A. calyculata

with prominent green spines and grayish "pencil" lines beneath; long-lasting inflorescence on erect panicle with brilliantly red bracts and yellow flowers.

A. brevicaullis (brev-ee-col-lis) is a thin, tubular plant to 14 inches (35.6 centimeters) with grayish-green leaves spotted maroon; spines on edges; lovely orange flower bracts, yellow flowers.

A. calyculata (kal-lick-ew-la-ta), growing to 20 inches (50.8 centimeters), has dark-green leaves. The compact flower head brings a butter-yellow color to the windowsill all through the winter. This one will do best with good light and some coolness, about 55°F (12.8°C) at night, in winter.

A. caudata var. variegata (kaw-da-ta, var-ee-ga-ta), growing to 36 inches (91.4 centimeters) tall, would be a good investment for its handsome foliage even if it never flowered. The leaves are yellow-white and green-striped. The flower head, yellow-orange, erect, and striking, usually appears in winter. This is one of the few difficult Aechmeas but a healthy specimen is worth the effort.

A. c. albo marginata (al-bo mar-gee-na-ta) is about 40 inches (1 meter) tall with decorative green leaves handsomely striped with vertical bands of yellow.

A. chantinii (chan-tin-ee-eye), to 36 inches (91.4 centimeters), the queen of the Aechmeas, is indeed handsome. Leaves are olive-green with silver bands. The large flower head is upright, a vivid red and yellow. Of medium size, this regal plant gives weeks of pleasure when it blooms in spring.

A. chantinii

A. c. 'Burgundy' is a 36-inch (91.4-centimeter) vase of burgundy-colored leaves spotted and blotched with silver; branched flower spike with vivid rose bracts; yellow flowers. Outstanding.

A. c. 'Silver Ghost' is an upright 40-inch (1-meter) rosette of handsome, pale, silvery green leaves banded with darker green; spike erect and branched, pale rose bracts and greenish-yellow flowers.

***A. distichantha var. schlumbergeri** (dis-ti-**kam**-tha, shlum-**berg**-er-i)* grows to 16 inches (40.6 centimeters) or more with stiff heavily spined leaves. The violet flowers and bright rose bracts make it a desirable plant. Needs little care and is not particular about temperature.

***A. fasciata** (fas-see-**a**-ta)*, 24 inches (61 centimeters) tall, sometimes called the urn plant, is a window gardener's delight. It produces a large pink inflorescence with dozens of tiny blue flowers in summer. Fine for the beginner, sure to bloom.

A. f. 'Silver King,' to 48 inches with spectacular foliage, has gray-green rosette boldly barred in dark color; pink flowers in crown.

A. f. variegata (var-ee-ga-ta), a variegated silver vase, grows to 24 to 30 inches (61 to 76.4 centimeters), with the center of the channeled green leaves attractively striped and banded with silver cross-banding.

A. filicaulis (fy-li-cau-lis) grows to 24 inches (61 centimeters) and has satiny green leaves and dancing white flowers and red bracts on a thin pendant spike that comes alive in the slightest breeze. This unusual and decorative Aechmea is well suited to basket culture.

A. 'Foster's Favorite' grows to 24 inches (61 centimeters), with wine-red foliage and dark-blue winter flowers followed by red berries. A small, inexpensive air plant readily accommodated at a window.

A. fulgens var. discolor (full-genz, dis-kol-or) is only 20 inches (50.8 centimeters) tall, with maroon-shaded green leaves. In spring, it produces dark-purple flowers followed by rose-colored berries that last for months. It is well suited to a limited space and will survive considerable neglect.

A. fulgens discolor

A. mertensii

A. luddemanniana *(lewd-e-man-ee-**a**-na)* grows to 40 inches (1 meter) with arching, spiny green-copper leaves, and straight red bracts, blue petals; an unusual bromeliad.

A. mertensii *(mer-**tens**-ee-eye),* the China berry, is an epiphytic open rosette with few green leaves to 24 inches (61 centimeters) long and covered with white scales, especially beneath, and having marginal spines. The slender stalk has rose bracts and the many-flowered inflorescence has yellow or red petals; fruit is blue berries holding color from November to April.

A. 'Meteor,' to 36 inches (91.4 centimeters), is a tubular rosette of pale-green leaves, edges slightly spined; has an exquisite inflorescence of fiery red bracts and flowers.

A. mooreana *(moor-ee-**an**-a),* a very showy bromeliad, to 28 inches (71.1 centimeters), has an inflorescence similar to *A. chantinii* but with lower bracts carmine-rose and upper, flattened bracts lime-green tipped with flame orange. Foliage is bronzy green.

A. mertensii **(closeup)**

A. nudicaulis variegata *(newd-ee-**col**-lis var-ee-**ga**-ta),* a very handsome plant, makes a 30-inch (76.2-centimeter) tubular vase of bright-green leaves lightly lined with yellow, leaf edges spiny. Floral bracts bright rose, yellow flowers.

A. orlandiana *(or-land-ee-an-a)* is a showy 18-inch (45.7-centimeter) rosette of bright yellow-green leaves with brown cross-banding and heavy black spines. The arching spike has salmon-colored bracts and ivory flowers. Needs more heat than most in genus. Bright light makes leaves colorful.

A. ornata *(or-na-ta)* forms a loose-leaved 20-inch (50.8-centimeter) rosette with erect, hard, gray-green leaves; the inflorescence is a dense cone-shaped crown with berrylike bracts and pale-red petals. Give full sun.

A. penduliflora *(pen-dul-i-flor-a)* is a 24-inch (61-centimeter) rosette of straplike green leaves that are pink in bright light. The inflorescence is somewhat pendant with small yellow petals followed by vivid blue berries. A decorative species.

A. pubescens *(pew-bess-kens)*, to 14 inches (35.6 centimeters), is decorative in bloom and afterward in April when the white berries turn blue. Requires more light than other Aechmeas.

A. quesnelia marmorata x A. chantinii

***A. quesnelia marmorata x A. chantinii** (kes-nel-ee-a mar-moor-a-ta, chan-tin-ee-eye)* is a hybrid with 36-inch (91.4-centimeter) tubular vase of gray-green leaves slightly spined; bright red floral bracts and greenish-yellow flowers.

***A. racinae** (ra-seen-ee),* to 14 inches (35.6 centimeters), has light-green leaves and yellow-and-black flowers on a pendant spike. The orange-red berries appear appropriately at Christmastime and last until April. Try this one in a hanging basket and grow it in light shade.

A. 'Rajah' is a tubular, pale-green, upright rosette to 30 inches (76.2 centimeters) with erect flower spike; pale orange bracts and yellow flowers.

***A. ramosa** (rah-moh-sa),* about 30 inches (76.2 centimeters) tall, has waxy green leaves and a large branched yellow flower head with rose-colored bracts followed by greenish berries. A handsome Aechmea that holds color for many months.

***A. recurvata var. ortgiesii** (re-cur-vat-a ort-gees-ee-eye)* is 20 inches (50.8 centimeters) tall with thorny cactuslike foliage and a low pink flower head. Grow it on the dry side and in sun.

A. recurvata ortgiesii

A. 'Red Wing'

A. 'Red Wing,' a stunning Hummel hybrid involving ***A. penduliflora x musica* (*mew-si-ca*)**, is a handsome 34-inch (86.4-centimeter) rosette with large coppery leaves, dark purple beneath. The wine-red stalk bears an inflorescence of many berries, first pink then purple, in heavy clusters; the flowers are straw-colored.

A. 'Spring Beauty,' to 30 inches (76.2 centimeters), is a handsome tubular rosette of pale olive-green leaves, edges spined; flower spike erect and short; rose bracts and yellowish flowers.

***A. tillandsioides* (*till-and-see-oy-deez*),** to 12 inches (30.5 centimeters), is an epiphytic rosette with narrow, leathery, grayish leaves armed with marginal spines; inflorescence has serrated floral bracts in green, yellowish, or red; flower petals yellow, followed by berries first white then blue.

***A. t. lutea** (lew-tee-a)* is an open rosette with narrow leathery gray-green leaves, slightly spiked; flower spike is erect with orange bracts and yellow flowers.

***A. weilbachii** (weel-**bach**-ee-eye)* grows to about 14 inches (35.6 centimeters), with green leaves and an almost pendant, branching, red-and-lavender inflorescence in late fall. Very decorative and a good grower.

***A. zebrina** (zee-**bry**-na),* a very elegant plant, is a 36-inch (91.4-centimeter), urn-shaped, dense rosette of olive green, stiff, broad leaves rounded at apex, and covered beneath by crossbands of white scales and thorny margins; branched inflorescence of spreading spikes with crimson bracts. The flat flowering branches are in a series of yellow boat-shaped bracts with yellow flowers. Unlike *A. chantinii, A. zebrina* has sepals covered by the bracts.

A. weilbachii

ANANAS (a-NAY-nus)

CULTURE

Water: Likes evenly moist potting medium; must have good drainage.

Potting medium: Fir bark or soil-and-bark combination.

Feeding: A few times in summer with 10-10-5 plant food.

Light: Bright sun.

Temperature: Average home temperature fine (60°F to 80°F/15.6°C to 26.7°C).

Propagation: By offshoots.

Remarks: With exception of *A. nanus*, most species very large and difficult to handle because of spines.

Ananas is the oldest known group of bromeliads. It includes the pineapple, which has been widely grown for several centuries. In the fifteenth century the pineapple was a known crop in the New World. As houseplants the several species have their merits but are not really exceptional because they simply are not as decorative as other bromeliads. The miniature variety seems to have found a place in American homes more as an oddity than as a good indoor subject.

There are about ten known species of Ananas, mostly terrestrial, native to Brazil, Paraguay, and Venezuela. They have been successfully cultivated in Hawaii for some years. Most species are large rosettes—to about 48 inches (1.2 meters) across—and thus demand space. Leaves are either plain green or, in the decorative species, striped yellow and green and are armed with spines.

The fruiting head develops on top of a stout stem that comes from the center of the plant; the head bears a small rosette of leaves that is a miniature version of the parent plant. This head can be cut and started as a new plant. The flowers have purple-blue petals.

I have grown several kinds of Ananas but did not find them that intriguing, except for *A. nanus*, which is a fine houseplant because it is small. Plants require good sunlight to grow well but can tolerate warmth or coolness (55°F/12.8°C).

A. bracteatus *(brack-tee-**a**-tus)* has 40-inch (1-meter) dark-green leaves that turn a handsome red in sunlight. Leaves are edged with spines, so be forewarned. A good bromeliad if you have the space.

A. b. variegatus *(var-ee-**ga**-tus)* is large, to 40 inches (1 meter) and has exquisite leaf coloring of pink, white, and pale green. Very dramatic.

A. comosus *(koh-**moh**-sus)* is the commercial pineapple plant, growing to 30 inches (76.2 centimeters) tall. The stiff, spiny-edged leaves are green and the purple flowers are produced at the top of an erect stalk.

A. nanus *(**nay**-nus)* is a 10-inch (25.4-centimeter) dwarf species with satiny green leaves 10 to 14 inches long (25.4 to 35.6 centimeters); it is available at florist shops and somewhat resembles *A. comosus*. The small pineapple is decorative on its stiff stalk.

ARAEOCOCCUS (air-ee-o-COCK-us)

CULTURE

Water: Keep potting medium evenly moist, never too dry or soggy. Mist plants occasionally to maintain good humidity.

Potting medium: Grow in fir bark or use equal parts bark and soil. Be sure drainage is good.

Feeding: Not necessary.

Light: Keep in bright light—no sun. Will also tolerate shaded places.

Temperature: Average home temperatures of 75°F (23.9°C) by day and ten degrees less at night is fine.

Propagation: By offshoots.

Remarks: Amenable indoor plants—require little care.

This is a genus of only a few known species; two are popular with hobbyists and make good indoor plants. Coming from Costa Rica, Colombia, and Venezuela, the plants are epiphytes and have thin, whiplike leaves growing in a graceful rosette.

The plants do best in fir bark with even moisture all year; they like a somewhat humid atmosphere. Good for spot decoration in the home on table or desk.

A. flagellifolius *(fla-gel-ee-**fol**-ee-us)* grows to 24 inches with narrow reddish terete foliage, pink flowers, and blue-black berries on a short scape. Of unusual appearance.

A. pectinatus *(pec-teen-a-tus)*, another medium grower, has reddish-bronze leaves and a red inflorescence.

BILLBERGIA (bil-BERG-ee-a)

There are some good plants in this group and some not so attractive ones. The *Billbergia nutans* varieties take up a lot of space, and the plain green foliage is not very pretty. However, these plants do bloom profusely—sometimes three times a year. Most Billbergias are native to eastern Brazil; however, a few come from Mexico and Central America. In nature the plants prefer an arboreal existence: I have seen trees literally covered with Billbergias in Central America. However, the plants are so adaptable that they also grow at ground level.

Billbergias have fewer leaves than most bromeliads: five to eight. They are easily distinguishable from other plants because they are definitely tall and tubular; they do not have the rosette shape of the Aechmeas or Guzmanias. Leaves can be plain green, but usually they are mottled, banded, or speckled with contrasting color. Most Billbergias have brilliantly colorful bracts (pink or rose) and as a group I find them more floriferous and easier to bring into bloom than most bromeliads. The flowers are generally small, of vivid purples, blues, and greens. The inflorescence is pendant and short-lived, perhaps three to five days.

Billbergias grow almost by themselves and need very little attention. If situated in a bright place and given even minimal watering, they grow—and fast. In fact, they grow so quickly that they need frequent repotting, which can become bothersome. I grow my plants in equal parts of sand, fir bark, and soil. Drainage must be perfect, and the growing medium should be kept just moist, with the vase filled often with water. Mist plants with tepid water frequently, especially during hot weather.

Bright locations produce very colorful foliage; in shade, foliage is somewhat dull and not as colorful. Even in minimal light the plants survive for a long time: I have one potted *B. zebrina* that has adjusted to a less-than-optimum situation in my living room, and blooms yearly.

These plants are favorites in the garden in all temperate regions, where they will naturalize on trees and create handsome scenes. I had Billbergias outdoors in the garden in Chicago, Illinois, until late October. (See Chapter 5 for more information about plants outdoors.) Billbergias produce offsets easily, or you can divide plants by slicing a crown.

Billbergias are not as demanding as most bromeliads in terms of air circulation and will, if necessary, survive in even steam-heated, unventilated areas.

B. amoena *(a-mee-na)*, growing to 28 inches (71.1 centimeters), is tubular with green leaves and rose-colored bracts. Flowers are green with blue edges.

B. a. var. viridis *(vir-eye-dus)*, a handsome, medium-size, tubular, 30-inch

CULTURE

Water: Keep some water in cup at all times. Keep potting medium somewhat dry, watering about once a week all year. Mist leaves when practical.

Potting medium: One part medium grade fir bark, one part sand or perlite, and one part soil.

Feeding: Fertilizer administered once a month to cup of plant.

Light: Will tolerate shade if necessary; best light is bright; direct sun not necessary.

Temperature: Average to warm; 80°F (26.7°C) by day, ten degrees less at night.

Propagation: From offshoots in spring.

Remarks: Many kinds available; not all are attractive so make selections with care.

(76.2-centimeter) plant with ivory-spotted green leaves and green petals; blossoms in April for me.

B. brasiliensis *(bra-zil-ee-**en**-sis)* is a tubular rosette, 36 inches tall, (91.4 centimeters) of broad, silver-banded leaves. Large rosy-red scapes with satin-blue petals appear in late summer.

B. distachia *(dis-**tack**-ee-a)*, small and bulbular, to 20 inches (50.8 centimeters); has green leaves with purple coloring. The cascading scape has pink bracts and green flowers tipped blue.

B. elegans *(**ell**-e-ganz)*, a vaselike rosette; has broad, light-green leaves margined with brown spines; grows to 16 inches (40.6 centimeters). Pendulous inflorescence with rose bracts and yellow-green flowers. Takes coolness (50°F/10°C).

B. euphemiae *(ew-**feem**-ee-ee)*, with gray-green leaves to 20 inches (50.8 centimeters), has bracts with dark-blue petals; blooms in any season. Not as showy as others in the genus.

B. 'Fantasia,' a 30-inch-tall (76.2-centimeter) hybrid, is a perfect size for the

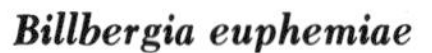

Billbergia euphemiae

B. 'Muriel Waterman'

windowsill with green leaves spotted ivory, green, or rose. The pendant inflorescence of scarlet bracts and blue flowers makes it most appealing.

***B. horrida** (hor-id-a)*, a few-leaved vase to 20 inches (50.8 centimeters), has heavily spined, gray-banded leaves and an erect spike of greenish petals tipped blue and circled by pink bracts. A very colorful medium-size plant.

***B. leptopoda** (lep-toe-po-da)* with broad green leaves spotted cream color, rarely grows above 12 inches (30.5 centimeters). Leaf tips curl and the inflorescence has pink bracts and green-and-blue flowers. A good pot plant.

***B. lietzei** (leetz-ee-eye)*, only 12 inches (30.5 centimeters) tall, is a charming small plant that lights your window with bright cerise flowers around the holiday season.

***B. meyeri** (my-er-eye)*, tall and thin, to 24 inches (61 centimeters), has gray leaves mottled with silver; produces a trailing pink scape with green-and-blue petals. Grow it on the dry side.

B. 'Muriel Waterman,' a popular bromeliad, grows to 30 inches (76.2 centimeters), with tubular shape; leaves plum-colored with silver-gray bands; pink bracts, vivid blue petals.

***B. nutans** (new-tanz)*, called Queen's Tears, is different in the genus. It is a

B. pyramidalis striata

B. splendens

tight, tubular vase to 30 inches (76.2 centimeters), with dark-green, narrow, arching leaves and nodding flower scapes; green tipped and violet blue. I grow a specimen plant in a big pot; it blazes with color at Christmas. This species grows fast and delights every plant lover.

***B. porteana** (por-**tee**-a-na)*, with gray-green leaves, is 36 inches (91.4 centimeters) tall with a cascading scape of pink bracts and green petals. A slow-growing but excellent species, often confused with *B. zebrina*.

B. pyramidalis* var. *concolor *(per-am-i-**dal**-is, kon-**kol**-or)* is a favorite with collectors. It is a broad-leafed 24-inch (61-centimeter) bottle-shaped plant,

B. zebrina

golden-green, the compact flower head densely set with pink to red flowers on a short scape. A very beautiful Billbergia that blooms in winter. *B. p. striata* (*stree-a-ta*) combines beautiful flowers with pale-green and white foliage.

B. sanderiana *(san-der-ee-**a**-na)* makes a medium-size, few-leafed vase. The foliage varies with light exposure from grayish-green almost to red. The nodding inflorescence is rose, green, and blue.

B. splendens *(**splen**-denz)* is a large rosette to 30 inches (76.2 centimeters) of pale-green leaves; handsome pink bracts on a pendant flower scape.

B. venezueleana *(ven-ez-way-lee-**an**-a)* is a big, tubular species growing to 40 inches (1 meter). The leaves are patterned with chocolate-brown and silver bands, a startling combination. The scape is equally bold, laden with broad, usually pink bracts and purple petals. Needs somewhat dry growing conditions.

B. vittata *(vy-**tat**-a)* is a tubular rosette about 36 inches (91.4 centimeters) high of olive-green, spiny leaves covered with gray coating. The arching inflorescence is showy and has bright-red bracts; petals are violet and green.

B. zebrina *(zee-**bry**-na)* is broad-leafed and tubular, to 36 inches (91.4 centimeters), with gray-green foliage banded with silver, large, showy pink bracts, and, as a rule, golden flowers in a trailing scape. A handsome Billbergia that puts on a fine show in summer or autumn, this species requires dry conditions. Also known as *Aechmea zebrina*.

BROMELIA (bro-MEEL-ia)

Members of this group are rarely seen in plant collections, and yet there are two species I have grown for years that I think are extremely handsome: *Bromelia balansae* and the dwarf *B. humilis*. Mine have been perfect houseplants, and they make a dramatic statement in bloom because the entire center of the plant turns fiery red, like a volcano. They always elicit comments from my friends.

There are actually about fifty species of Bromelias, and I think more will be introduced into culture in the future. The plants are native mainly to Brazil, Argentina, and Paraguay, but some are found in Mexico, Guatemala, and the Caribbean. In nature Bromelias grow in thickets on the ground and are often used in place of hedges to define property and keep out intruders.

With the exception of *B. humilis*, most are very large plants, to 5 feet (1.5 meters) or more across, growing in a rosette shape, with plain green leaves edged with sharp spines—difficult to handle as a houseplant, but no more so than cacti. When potting or repotting Bromelias, handle them with newspapers or burlap to avoid being scratched. If you are scratched by the spines, do treat scratches or they may become infected. A few Bromelias have variegated foliage. The plants I have do much better in hanging baskets than in clay pots.

Indoors, Bromelias have little trouble adjusting to varying conditions and so can be grown in warmth or coolness (to 50°F/10°C), in bright light

CULTURE

Water: Keep potting medium evenly moist; can tolerate dryness if necessary.

Potting medium: Large-grade fir bark.

Feeding: Not necessary.

Light: Likes sun.

Temperature: Likes it hot, 80°F (26.7°C), ten degrees less at night; but will tolerate some coolness.

Propagation: By offshoots.

Remarks: Smaller species can make handsome indoor display; tough to handle because of spines on leaves.

or in shade. I grow my plants in large-grade fir bark with no soil. Keep drainage perfect, and water plants frequently and heavily. I try to give my plants a very sunny place because then the foliage becomes highly colored. I have found that good direct sun is necessary if you want plants to produce their fiery display of color in midsummer. The plants produce many offsets.

Bromelia foliage is so tough that insects never come near them, and I have never had any disease strike my plants. As decorative plants, Bromelias are especially good when, as mentioned, they hang in baskets so their full beauty can be seen.

B. balansae (bal-an-see) has spiny, hooked, dark-green leaves 4 to 5 feet (1.2 to 2.1 meters) long with red bracts and a cone-shaped inflorescence of bluish-white flowers. At blooming time the center erupts into a volcano of flame red. A splendid sight.

B. humilis (hew-mil-is) is smaller, about 2 feet (61 centimeters) across, with narrow-toothed green leaves. The red heart of flame surrounds hidden flowers in the core of the plant.

B. serra variegata (sar-ah var-ee-ga-ta) is a 48-inch (1.2-meter) rosette of spiny-edged brown-gray leaves; center of plant turns fiery red at bloom time. Spectacular, but only for those who have space for it. Handle with gloves.

CANISTRUM (can-IS-trum)

I had been growing bromeliads for about two years when I visited the Fantastic Gardens in Miami, Florida, and saw my first Canistrums. To me the plants resembled Neoregelias and Nidulariums but were somewhat more upright in growth and a little smaller. I shipped three plants home and found they made excellent houseplants.

Canistrum is a small genus of only six species native to Brazil. They have rosette-type growth and pale-green leaves usually mottled with darker green. The white and green flowers are small and the flower head compact, hidden in the center of the plant surrounded by colorful bracts that hold the blooms. In nature Canistrums prefer arboreal locations, usually growing on tree limbs or on rocks. Plants vary from 20 to 30 inches (50.8 to 76.2 centimeters) across and make fine desk or window plants.

I grow Canistrums in a bright place. Sun is not absolutely necessary—indeed, they seem to do better in sheltered light. They like moisture and need somewhat more humidity than their cousins the Neoregelias. Keep leaves clean—I wipe the foliage with a damp cloth—and keep the humidity at about 30 percent. My plants have not been as free with offsets as most other bromeliads and generally have been more difficult to bring to bloom than other plants in the Bromeliaceae.

C. cyathiforme (sigh-ath-ee-form-ee) is a handsome rosette to 30 inches (76.2 centimeters) with bright-green leaves spotted brown; flower crown has reddish bracts and yellow petals.

C. fosterianum (fos-ter-ee-an-um) grows to 16 inches (40.6 centimeters) with grayish-green mottled leaves. The large flower head is raised and surrounded with brilliant pink bracts; a very handsome species.

CULTURE

Water: Keep water in cup at all times and carry potting medium slightly moist—more moist than most bromeliads.

Potting medium: One part fir bark, one part perlite, one part soil.

Feeding: Not necessary.

Light: Low to moderate light; sun not necessary and direct rays can harm.

Temperature: Somewhat warm, say, 80°F (26.7°C), ten degrees less at night.

Propagation: From offshoots at base.

Remarks: Very tough plants and can take abuse if necessary. Highly recommended.

Canistrum lindenii

C. lindenii *(lin-den-ee-eye)* is 28 inches (71.1 centimeters) across with light-green leaves spotted dark green. The rose-shaped inflorescence has white bracts and is either sunk in the cup or rises well above it. Usually summer flowering.

C. l. albo marginata *(al-bo mar-gin-a-ta)* is a 26-inch (66-centimeter) rosette with exquisite yellow and green banded leaves. Rose-shaped inflorescence has whitish-green bracts.

C. l. var. roseum *(ro-zee-ahm)*, 28 inches (71.1 centimeters) across, has leaves flushed pale pink with rose-colored bracts and white and green flowers.

CATOPSIS (ca-TOP-sis)

I think these are handsome plants, ideal for indoor culture. They are generally small, tubular, bear beautiful but bizarre flowers, and grow very well with only bright light and occasional watering.

There are about twenty-five species found in Mexico, Central America, and the West Indies; a few grow in South America. These epiphytic plants generally have smooth-edged, waxy leaves that are solid green with undersides coated with white powder. Some species have banded or speckled foliage. Plants bear a branched inflorescence that may be pendant or somewhat erect. The small flowers have white or yellow petals.

I grow Catopsises in fir bark in small pots and water them frequently. I always keep the vases filled with water, and mist plants during hot weather. A bright place is the only other requirement for these plants to thrive. In fact, Catopsises are so undemanding that there is little more to be said about cultivating them. They make unusual accents for desks or small tables and will even grow away from light for many months.

***C. berteroniana** (ber-ter-own-ee-**an**-a),* to 24 inches (61 centimeters), has apple-green leaves, green bracts, and small white flowers that usually appear in the spring.

CULTURE

Water: Keep cups of plants filled with water at all times. Keep potting mix or bark just barely moist.

Potting medium: Bark, or soil-and-bark combination.

Feeding: Not necessary.

Light: Bright light, or will tolerate north exposure if necessary.

Temperature: Average home temperatures (60°F to 80°F/15.6°C to 26.7°C).

Propagation: Offshoots.

Remarks: Good small plants for indoor accent; easy to grow.

C. l. roseum

Catopsis sessiliflora zebrina

C. floribunda *(flor-ee-**bund**-a)* grows to 14 inches (35.6 centimeters), with a bottle shape, bright-green foliage, and pretty white flowers on a tall, arching spike that appears in spring. Excellent for the beginner.

C. morreniana *(moor-en-ee-**an**-a),* a small, 10-inch (25.4-centimeter) vase has gray-green foliage with a branched inflorescence. A pretty species.

C. nutans *(**newt**-anz)* grows only 7 inches (17.8 centimeters) high with broad, pointed leaves. The small flowers are bright yellow.

C. sessiliflora zebrina *(ses-si-**floor**-a ze-**bry**-na),* to 20 inches (50.8 centimeters), has a funnel-like rosette of dark-green leaves banded with darker color. Simple inflorescence rises above foliage; white petals.

CRYPTANTHUS (cript-ANTH-us)

Cryptanthuses are foliage plants—the inflorescence is insignificant, but the leaves are exquisite and as handsome as the popular Rex begonias and much easier to grow. A window can accommodate as many as twelve to fifteen plants because most Cryptanthuses are small, growing to only about 10 inches (25.4 centimeters) across. As houseplants they are amenable and adjust to conditions, but it takes some time before they grow readily.

There are about fifty species in the group and dozens of hybrids. These terrestrial plants come from Brazil and in nature they grow under varied conditions, but in cultivation they prefer a warm and bright location. Most species are low-spreading rosettes, with about ten to twelve 2- to 12-inch (5.1- to 30.5-centimeter) leaves per plant. The leaves are somewhat crinkled and generally mottled and striped with unusual colors: brown, rose, silver, chartreuse, copper, pink, and so on. All species have tiny white flowers in the center of the plant. Offshoots are produced between the leaves or by stolons and multiply readily.

As a group the Cryptanthuses have puzzled me. I have tried various ways of growing them but have not found any one way better than another. At one time I grew them in a terrarium; another time I wired them to a large tree branch and put them in direct sun at a window. Still another time I used them as decorative plants in small matching containers on my desk under artificial light. Either my test periods were not long enough or I did not have a control group, but I never found that plants responded better one way or another—that is, with more colorful leaves or more flowers. I decided to grow Cryptanthuses on a tree branch, and that is where they have been for more than ten years. The roots are imbedded in years-old osmunda, and the plants look attractive. The branch is suspended from a wire in the ceiling so plants are at eye level.

You can also grow the plants in standard houseplant soil in 2- or 4-inch (5.1- or 10.2-centimeter) pots. Growth is slow. Insects rarely bother the

CULTURE

Water: Let potting medium dry out between waterings. Do not wet leaves.

Potting medium: One part soil, one part fir bark, one part perlite.

Feeding: Give 10-10-5 plant food once a month to potting medium.

Light: Prefer only bright light; sun not necessary.

Temperature: 80°F (26.7°C) by day, ten degrees less at night.

Propagation: Offshoots are borne in leaf axils and can be removed and potted separately.

Remarks: Make excellent terrarium plants.

Cryptanthus acaulis

C. 'Aloha'

plants. I recommend these bromeliads if you have limited space and want something other than philodendrons or grape ivy. Keep humidity at about 30 percent, and mist foliage occasionally. *Cryptanthus* species under artificial light look spectacular and grow readily.

C. acaulis *(a-kaw-lis)* grows to about 5 inches (12.7 centimeters) across; the pointed leaves are apple green with a slight gray overcast. Tiny white flowers hide deep in the center of the plant.

C. 'Aloha' is about 10 inches (25.4 centimeters) across with typical multicolored leaves and bright white flowers. Leaves with scalloped edges.

C. beuckeri (bew-ker-eye) is about 10 inches across, with green and cream spoon-shaped leaves.

C. bivittatus (by-vi-ta-tus) sometimes sold as ***C. roseus pictus (ro-zee-us pictus),*** is perhaps the most common, frequently used in dish gardens because of the spectacular salmon-rose and olive-green leaves.

C. bromelioides (bro-meel-ee-oy-deez) grows 12 to 14 inches (30.5 to 35.6 centimeters) high. Leaves are copper red in sun, gray green in shade. Very easy to grow.

C. b. var. tricolor (tree-kol-or) is my favorite in the genus. It is larger than most species, with pointed upright, highly colored leaves about 12 inches (30.5 centimeters) long. They are striped and lined lengthwise with white, rose, and olive green. Sometimes called the Rainbow Plant.

C. 'Bueno Funcion'

C. 'Koko'

C. 'Minibel'

C. 'Bueno Funcion,' to 10 inches (25.4 centimeters) across, has somewhat wide, dark green-gray foliage and handsome small white flowers.

C. fosterianus *(fos-ter-ee-**a**-nus)*, about 20 inches (50.8 centimeters) across, has thick, stiff brown leaves with zigzag crossbands.

C. f. 'Elaine' is a hybrid with leaf color more pronounced.

C. 'It' to 10 inches (25.4 centimeters) across is popular because it has tricolor foliage: pink, green and brown. Very pretty.

C. 'Koko' grows to 8 inches (20.3 centimeters) across with grassy-green leaves and slightly wavy white flowers.

C. 'Minibel' forms a 14-inch (35.6-centimeter) rosette of rose-colored leaves edged dark gray-green; edges spined. White flowers.

C. tricolor *(tree-**kol**-or)* is a beautiful 10-inch (25.4-centimeter) rosette of green, white, and rose leaves; small white flowers.

C. zonatus *(zoh-**na**-tus)* has broad, wavy brownish-green leaves to 14 inches, (35.6 centimeters), crossbanded with irregular silver markings. The white flowers are hidden in the leaf axils.

DYCKIA (DICK-ee-a)

Dyckias, which resemble cactus, are ideal houseplants for the beginning gardener because they are very adaptable to most growing conditions and can, if necessary, withstand drought. The plants range in size from 6 inches (15.2 centimeters) to about 48 inches (1.2 meters) in diameter and have stiff, spiny-edged leaves. The smaller species make handsome desk and table accents in decorative pots.

I have grown Dyckias in fir bark and in soil; they seem to do best in bark. Originally from Central Brazil, these are rock-growing xerophytes growing in full sun in rocky crevices. The plants need excellent drainage to thrive—any waterlogged medium harms plants. Grow them somewhat dry at all times. They have excellent leaf coloring in sun but even in shade will not die although foliage color is less pronounced.

Because of the spines, plants are difficult to handle; wear gloves. Once potted, Dyckias require no more repotting for several years. In fact, I have grown some plants in the same container for over five years.

D. brevifolia *(bre-vee-**fol**-ee-a)*, once known as ***D. sulphurea*** *(sul-**fure**-ee-a)*, grows 10 to 15 inches (25.4 to 38.1 centimeters) across with stiff, succulent dark-green leaves. The tall spray of orange flowers appears at various times. Will do wonderfully in the garden and can take some light frost.

D. fosteriana *(fos-ter-ee-**a**-na)* is the most popular species and rightly so. It makes a handsome 12-inch (30.5-centimeter) plant with silvery purple rosettes that cascade over the pot, recurve, and form a fountain of leaves. Large orange flowers appear in spring or summer.

D. frigida *(**frig**-ee-da)*, to 30 inches (76.2 centimeters), has spiny, recurved leaves. The narrow stalk is branched and produces small but very pretty orange flowers.

CULTURE

Water: Allow potting medium to dry out between waterings; keep somewhat dry in winter.

Potting medium: One part gravel, one part sand, one part soil or fir bark.

Feeding: Use 10-10-5 once a month all year.

Light: Bright sun.

Temperature: Not fussy; anything from 60°F to 90°F (15.6°C to 32.2°C).

Propagation: Offshoots at springtime.

Remarks: Excellent and tough plants, rarely subject to any ailments. Grow slowly but can be pretty. Use caution when handling; wear gloves as most have spiny leaves.

D. 'Lad Cutak,' a vigorous grower, has a stiff ornamental rosette to 12 inches (30.5 centimeters) of spreading leaves, bronzy green and concave above, green or grayish-scaly beneath; spiny at margins; has a tall, erect, attractive inflorescence with orange flowers.

D. leptostachya *(lep-to-**stack**-ee-a)* is 20 inches (50.8 centimeters) tall with reddish-brown foliage and attractive orange flowers.

D. marnier lapostollei *(mar-nee-air la-post-**ol**-ee-eye)*, a rare plant, is 10 inches (25.4 centimeters) across with gray-green leaves, edges spined. Nice compact shape.

D. rariflora *(rar-ee-**floor**-a)* grows to 14 inches (35.6 centimeters) with silvery-green foliage and orange flowers.

***Dyckia* 'Lad Cutak'**

D. marnier lapostollei

FASCICULARIA (Fa-see-you-LA-ree-a)

This is a genus of only a few plants; they have dense rosettes of narrow spiny leaves and their beauty is at bloom time when the center of the plant lights up like red neon, making a brilliant show. They are from Chile, where they grow high in the mountains, either on rocks or in soil, and can tolerate quite cool temperatures (to 45°F/7°C).

Fascicularias are best grown in large tubs in a sandy soil; repot only every fourth or fifth year as the plants do not like to be disturbed. They revel in full sun: the picture we show was taken in the outdoor area at the University of California Botanical Gardens.

F. pitcairnifolia *(pit-care-nee-fol-ee-a)* is a spiny, dark-green, 30-inch (76.2-centimeter) rosette. At blooming, the center turns fiery red, making a dramatic statement. Plant likes heat and sun along with cool nights.

CULTURE

Water: Likes plenty of water but must have excellent drainage.

Potting medium: Use a sandy soil that drains readily.

Feeding: Not necessary.

Light: Likes plenty of sun.

Temperature: Will tolerate cool (45°F/7°C) nights but needs daytime heat.

Propagation: By offshoots.

Remarks: A very stunning plant in bloom; requires large container. Difficult to handle because of spines.

GUZMANIA (guz-MAY-nee-a)

CULTURE

Water: Keep cups of plants filled with water and potting mix evenly moist all year. Likes water.

Potting medium: One part soil, one part gravel, perlite, or fir bark.

Feeding: Weak fertilizer once a month in warm weather; not at all rest of year.

Light: Moderate to low light. No sun.

Temperature: Quite warm, to 80°F (26.7°C) by day; ten degrees less at night. Never below 55°F (12.8°C).

Propagation: Offshoots at base.

Remarks: Highly decorative plants but need more attention than most bromeliads. Potting medium must never be dry and leaves must be washed frequently.

These small- to medium-sized bromeliads almost outrank Aechmeas as the best bromeliads, and rightly so because there is an incredible variation of leaf color and pattern available within Guzmanias. The decorative-leaved beauties, such as *G. lindenii* provide startling room accent, and even seedling plants are colorful.

With about 135 species, Guzmanias grow in the forests of Ecuador and Colombia; other species come from Central America, Brazil, Costa Rica, and Panama. The plants grow in varying elevations, and it is not unusual to find Guzmanias at over 8,000 feet where nights get very cold. Generally, Guzmanias are in shady situations atop tree limbs but protected from direct sun—they grow on lower limbs rather than treetops. Being mainly epiphytic, the plants resent soil at the roots, although a few do grow as terrestrials in nature. Plants have smooth-edged, glossy green leaves or leaves with stripes, bands, spots, or crossbands in varying colors—indeed, there is a tapestry of color here. The basic growth pattern is a rosette, and most plants are about 3 inches (7.6 centimeters) across, making them fine for indoor space. The flower spike is erect and bears brilliantly colored bracts, from yellow to flaming red and burnt orange. The bracts last for some time. The tiny flowers are white or yellow.

My Guzmania collection at an east window grows very well and produces beautiful color. They are in clay pots in fir bark. I flood the plants almost daily in very hot weather but do not water much in cooler weather. In any kind of weather I always keep the vase filled with water. Do not feed these plants. The Guzmanias I fed reacted poorly—leaf burn can result. I mist plants frequently with tepid water.

Unlike most bromeliads, Guzmanias are not free with offshoots, producing only a few "*kikis*," and are generally slow growing compared to the Billbergias. There are small plants, medium growers, and a few large ones to 48 inches (1.2 meters) across. Here is an incredible array of good indoor plants for all to try. Pests rarely bother these bromeliads.

G. berteroniana *(bert-er-o-nee-**a**-na)* is a 24-inch (61-centimeter) rosette of shiny green leaves. The showy inflorescence has orange-red bracts with yellow flowers. A fine plant for the windowsill.

G. lindenii *(lin-**den**-ee-eye)*, a 40-inch (1-meter) rosette, has spectacular green foliage marked with transverse wavy lines of dark green, red beneath. The erect flower scape has green bracts and white flowers. Likes coolness.

G. lingulata *(ling-ew-**la**-ta)* is a handsome species about 18 inches (45.7 centimeters) across with a star-shaped, orange flower head.

G. l. major, growing to 30 inches (76.2 centimeters) across, has a larger flower.

G. l. var. minor, to 24 inches (61 centimeters) across, has a red-orange inflorescence with white flowers.

G. 'Magnifica' (*lingulata x minor*) is a hybrid, 30 inches (76.2 centimeters)

Guzmania lingulata major

G. 'Meyer's Favorite'

G. 'Minnie Exodus' (closeup)

across and a good one for the beginner. Its graceful rosette of leaves is decorative and the red flower head is star-shaped and forms lower in the crown than in most species of the group.

G. 'Meyer's Favorite' grows to about 24 inches (61 centimeters) across with green leaves tinged red; flower crown is red. Small version of G. 'Magnifica.' Easy to grow.

G. 'Minnie Exodus' makes a compact 24-inch (61-centimeter) rosette of green leaves with a slightly raised, fiery red flower crown with yellow flowers. Superlative houseplant.

G. monostachia *(mon-o-**stack**-ee-a)* ***(G. monstachya)***, about 24 inches (61 centimeters) across, has satiny green leaves arranged in a dense rosette. The pokerlike flower spike is erect with white flowers and green bracts stenciled with maroon lines. The very tip of the inflorescence is crowned blood-red in many varieties, orange in others. A very showy Bromeliad.

G. musaica *(mew-**say**-eek-a)* is sure to please houseplant enthusiasts. The leaves are 24 inches (61 centimeters) long, bright green, banded and overlayed with irregular lines of dark green and wavy purple markings on the reverse. The flower spike is erect and turns red at flowering time. The white waxy flowers are set tight into the poker-shaped flower head.

G. 'Orangeade,' typical of the *G. lingulata* group, has a loose 30-inch (76.2-centimeter) rosette of green leaves and an exquisite, brilliant-red flower crown. Very pretty.

G. 'Symphonie' is a 30-inch (76.2-centimeter) rosette of dark-coppery fo-

G. monostachia

G. 'Symphonie'

liage; from the center rises the showy, starry head of pointed, lacquered crimson bracts with yellow flowers.

G. vittata *(vit-**ta**-ta),* to 28 inches (71.1 centimeters) across, has graceful, pointed, soft-green leaves barred with maroon on the reverse. The handsome foliage makes it valuable for every collection but it also produces a round, white flower head with greenish bracts edged purple. Even at the darkest window, this fine plant retains decorative leaves. Don't grow below 60°F (15.6°C).

G. zahnii *(**zahn**-ee-eye)* grows to 20 inches (50.8 centimeters), the delicate

G. zahnii variegata

leaves almost transparent and striped red-brown. The flower spike has bright-red bracts and white flowers that hold color for six to eight weeks.

G. z. variegata *(var-ee-**ga**-ta)* is a 20-inch (50.8-centimeter) plant with dark-green leaves beautifully striped with whitish-yellow.

HECHTIA (HECT-ee-a)

I have never been overly fond of Hechtias although many people like them; they are perhaps more bizarre than beautiful. Plants generally have spiny leaves resembling cactus and range in size from 6 inches (15.2 centimeters)

CULTURE

Water: Keep potting medium somewhat dry; reacts poorly to too much water. Do not mist.

Potting medium: Soil or combination of equal parts soil and bark.

Feeding: Not necessary.

Light: Bright sun; will not grow in shaded areas.

Temperature: Can tolerate heat or coolness (to 55°F/10°C).

Propagation: By offshoot.

Remarks: Excellent small plants for those who can't grow anything.

to 3 feet (91.4 centimeters). The plants coming from hot areas of Texas, Mexico, and northern Central America are terrestrial and grow in desert hillsides baking in sun, where leaves turn exquisite colors.

In cultivation they can, if necessary, withstand cool places (to 50°F/10°C) and still not suffer, making them fine additions to the cool home. The flowers are on branched spikes that come from the side of the rosette rather than from the center, as in the case of most bromeliads.

Grow Hechtias in a sandy soil and allow the soil to be somewhat dry—overwatering can cause harm. Keep them in bright light or in sun if you want the bronzy-red coloring that is so attractive. Repot only when necessary—these plants resent root disturbance.

H. argentea *(ar-gen-tee-a)* has large recurved silver leaves to 10 inches (25.4 centimeters) growing in a symmetrical rosette with a tall spike of orange flowers. A good showy Bromeliad.

H. glomerata *(glow-mer-a-ta)* is a small to medium plant with fleshy, green, recurved leaves armed with spines.

Hechtia glomerata

H. montana

H. montana *(mon-tan-a)* forms a 12-inch (30.5-centimeter) rosette of spiny silver-gray leaves; sprawling growth; leaves heavily armed with spines. Flowers greenish-white.

H. rosea *(ros-ee-a)* is very decorative with 24-inch (61-centimeter) rosettes of sharp saw-toothed green leaves.

H. texensis *(tex-en-sis)* with 6-inch (15.2-centimeter) rosettes of sharp saw-toothed green leaves, is a small grower and best suited for a dish garden.

HOHENBERGIA (ho-en-BERG-ee-a)

These large attractive plants, to 5 feet (1.5 meters), are not often seen, but I recommend them if you have the space. They do very well in a greenhouse

CULTURE

Water: Let potting mix dry out between waterings; keep vase of plant filled with water.

Potting medium: One part fir bark, one part soil.

Feeding: Not necessary, or once a month in summer.

Light: Will tolerate low light but does like some sun.

Temperature: Average, to 80°F (26.7°C) by day; ten degrees less at night.

Propagation: From offsets.

Remarks: Big plants but make stunning room decoration; grow in ornamental tubs.

and require minimal culture. I use *Hohenbergia stellata* in my plant room as an accent where it always causes comments in summer when it blooms; the large flower spike is vividly colored.

Hohenbergias dot the island of Jamaica. When I was there I saw several splendid specimens at the botanical gardens. They also come from Brazil. Plants are large, dense rosettes with pale-green leaves edged with tiny spines not unlike cactus spines. The pendant flower spike is long and arises from the center of the plant. The bracts are vividly colored, green to red. The small flowers are blue or white. Most Hohenbergias are epiphytic and indoors require large, heavy containers to keep them from toppling over.

I grow Hohenbergias in bright sun to produce handsome leaves (they do not do well in shade). I use a growing medium of equal parts fir bark and soil that drains readily. I water the plants heavily almost all year, except in the winter, when I give them a slight rest with less water. I feed plants twice a year, in spring and summer, and I always keep the vase filled with water. Hohenbergias like a warm spot (75°F/23.9°C), and frequent misting of the foliage is a good idea. Plants produce offsets but not in abundance. To start *kikis,* root them first in vermiculite for a few weeks or until roots show. Insects rarely attack these plants.

H. ridleyi *(rid-lee-eye)* grows to 5 feet (1.5 meters) with golden-yellow leaves and a tall, branched red-and-purple inflorescence.

H. stellata has 3- to 5-foot (.9- to 1.5-meter) spiny leaves and a tall spike with red bracts and purple flowers. An outstanding sight in any garden.

NEOREGELIA (nee-o-ree-GEEL-ee-a)

In Germany and Japan, Neoregelias are favorites indoors because they are handsome, undemanding, and among the finest foliage plants available. There are about sixty known Neoregelias, most native to eastern Brazil; however, Colombia and Peru also have a share of the plants. In nature the plants grow near the ground or in lower branches of trees, preferring a shady place with good air circulation.

Most plants are medium-sized and grow in a compact, flat rosette. Some are only a few inches across; others grow to 5 feet (1.5 meters). The brilliant foliage is plain green, spotted, marbled, striped, or banded with color, and at bloom time most plants show bright red to rose at the center of the plant. The flowers are small, usually in shades of blue or a combination of blue and white. Flowers die quickly, but the flush of color in the foliage remains for many months. Neoregelias are fine for north exposures and excellent if you have little time to spend but still want some colorful plants in the home.

I grow Neoregelias at a west exposure, where they get enough light but not too much sun and they thrive. I purposely keep them at ground level so their beautiful foliage can be easily seen. I use a potting mix of one-third perlite, one-third soil, and one-third fine-grade fir bark that drains readily. I flood plants with water in the warm months, but they are given less moisture in cool weather. I never feed Neoregelias, but I always keep the vase filled with water. In my plant room small insects and frogs live in the center of the plants—those insects that die in the plant are excellent nutrition for the frogs

Neoregelia in container

and the plant. Unfortunately, the water reservoir can attract mosquitoes. Occasionally wipe leaves with a damp cloth and spray foliage with tepid water, especially in the hot weather.

Neoregelias are rarely bothered by insects and are free with their offshoots, as many as four or five kikis. Plants become overcrowded in pots if you do not cut away the offsets. I purposely let this happen in a few pots to get a big statement of color.

N. ampullacea *(am-pull-ace-ee-a),* about 9 inches (22.9 centimeters) across, has leaves with mahogany crossbands and small blue flowers in spring or summer. Give a little more light than for most Neoregelias.

N. 'Bonfire' grows to 20 inches (50.8 centimeters) in diameter—a beautiful rosette of reddish-plum leaves. Flower crown deep in plant; tiny violet petals.

N. carolinae *(car-o-lye-nee)* is perhaps the showiest in the genus, the tapered leaves dark green, the rosette about 30 inches (76.2 centimeters) across; the center of the plant turns red before blooming. My plant was in full color for nine months. Undoubtedly one of the finest houseplants obtainable.

N. c. 'Meyendorfii' is a broad 30-inch (76.2-centimeter) rosette of flat olive-green lèaves with coppery tinting. At flowering time the inner leaves turn a dark maroon; flowers are lilac and deep in the center.

N. c. var. tricolor *(tri-kol-or)* is expensive but a panorama of color; the variegated leaves are white-striped. When in flower the foliage has a pinkish hue and the heart of the plant turns cerise. A highly desirable 30-inch (76.2-centimeter) bromeliad that steals the show. Grow it in shade with warmth.

CULTURE

Water: Keep water in cup at all times; allow potting medium to dry out between waterings.

Potting medium: One part soil, one part fir bark, one part perlite. Good drainage essential.

Feeding: None necessary

Light: Bright light or some sun.

Temperature: Likes warmth; to 80°F (26.7°C) by day; ten degrees less at night.

Propagation: Throws offshoots in abundance; remove when 2 to 3 inches (5.1 to 7.6 centimeters) tall and pot separately.

Remarks: Excellent table and desk plants; very decorative.

Neoregelia concentrica

N. compacta *(com-**pac**-ta),* is a dense, erect rosette to 24 inches (61 centimeters) of green leaves; has inner leaves that turn red at bloom time. Good small plant.

N. concentrica *(con-sent-**reek**-a),* about 30 inches (76.2 centimeters) across, has pale-green leaves flecked with purple and edged with black spines. Leaf tips are red. Before bloom the core of the plant turns fiery red-purple and tiny flowers are blue.

N. cruenta *(krew-**en**-ta)* forms a 24-inch (61-centimeter) upright rosette with straw-colored leaves edged with red spines. A most unusual *Neoregelia* that needs full sun.

N. johannis *(joe-**han**-nis)* is a durable 20-inch (50.8-centimeter) species with a lavender center. Although not as handsome as others in the genus, it is still worthwhile.

N. 'Marmorata' is a hybrid with yellow-green and crimson leaves; rosette to 30 inches (76.2 centimeters) across. Spring or winter flowering, it needs good light for proper leaf color. Flowers are white and deep in the cup. This one seems to thrive on neglect. It's for people who "can't grow anything."

N. cruenta

N. 'Painted Lady'

N. zonata

N. 'Painted Lady' forms a 24-inch (61-centimeter) rosette of dark-green leaves suffused with brownish-red markings; violet flowers.

N. 'Purple Passion' grows to 24 inches (61 centimeters) across; fine purple-red leaves; small pink flowers deep in cup.

N. 'Red Knight' makes a 20-inch (50.8-centimeter) rosette of handsome bright-green leaves heavily banded with maroon; flowers almost violet.

N. spectabilis *(spec-tab-i-lis)*, the Painted Fingernail plant, grows to 30 inches (76.2 centimeters) across with spine-free, leathery green leaves tipped cerise. The small blue flowers usually appear in warm weather.

N. zonata *(zoh-na-ta)* is a 20-inch (50.8-centimeter) tubular vase with pale-green leaves that are marked and banded with purple, edged with spines. Deep-blue petaled flowers nestle in cup of plant at bloom time.

NIDULARIUM (nid-you-LAR-ee-um)

Similar to Neoregelias, Nidulariums make excellent houseplants if you have limited time to devote to plants. They grow readily, produce handsome leaf color, and are usually small- to medium-sized plants suitable for almost any place in the home. They like more light than Neoregelias but otherwise need the same conditions.

The thirty-five species of Nidulariums are from eastern Brazil. They are flat rosettes with a fan shape and at bloom time the center of the plant turns red. Flowers are usually hidden in the center and are inconspicuous, usually red, white, or pale blue. Leaves are glossy and finely toothed; they vary in color from green to purple and may be striped or mottled.

I grow Nidulariums at a southwest window where they get excellent light; thus the foliage color is always handsome. I have plants in equal parts of perlite, fir bark, and soil that drains readily. I wash the leaves occasionally with a damp cloth and mist plants daily in hot weather. I keep the center of the plant always filled with water; I do not feed Nidulariums. Plants set offshoots in abundance, and the plants require very little care once they are established, almost growing by themselves. Insects are never a problem.

Grouped with Neoregelias, these plants make a handsome display; there are many fine new hybrids available.

***N. billbergioides** (bill-berg-ee-oy-deez)* is a small rosette to 30 inches (76.2 centimeters) and one of the few species that produces a flower head on a 6- to 8-inch (15.2- to 20.3-centimeter) stalk. The leaves are green, the bracts orange; tiny white flowers appear in summer or fall.

***N. b. var. citrinum** (si-tree-num)* has a bright-yellow secondary rosette.

***N. fulgens** (ful-genz)*, with pale-green leaves spotted dark green, is a 20-

CULTURE

Water: Water cup of plant all the time; keep potting mix barely moist all year, never dry or soggy.

potting medium: One part fir bark, one part soil, one part perlite; must have excellent drainage.

Feeding: Apply weak fertilizer once a month to potting medium in warm weather only (optional).

Light: Low to bright light; will also take shade.

Temperature: Can take some coolness. To 75°F (23.9°C) by day; ten degrees less at night.

Propagation: Produces many offshoots.

Remarks: Very amenable plants; excellent table or desk decoration.

Nidularium fulgens

inch (50.8-centimeter) rosette with a bright-cerise leaf cup at blooming time, which is usually in spring. A very decorative plant.

N. innocentii *(in-o-**sen**-tee-eye)* makes a 24-inch (61-centimeter) rosette with almost purple leaves. The white flowers nestle deep in a rusty-red cup.

N. i. lindenii *(lin-**den**-ee-eye)* is a 24-inch (61-centimeter) rosette of striped green-and-white leaves; center of plant turns fiery red at bloom time. Outstanding.

N. i. var. lineatum *(lin-ee-**a**-tum)* has creamy white-and-green-striped leaves, a 24-inch (61-centimeter) rosette with a flaming red center and white flowers in summer. This species is somewhat temperamental and requires warmth.

N. i. var. wittmackianum *(wit-mack-ee-**an**-um)*, smaller, has green leaves and small white flowers in winter or spring.

N. procerum *(pro-**see**-rum)* is a big rosette of 36 inches (91.4 centimeters) with long yellow-green leaves. The bright-red cup holds orange-red flowers; a very handsome color combination.

N. regelioides *(re-gel-ee-**oy**-deez)*, one of the first bromeliads I bought, makes a compact 14-inch (35.6-centimeter) rosette with dark-green leathery leaves. The flowers are red.

ORTHOPHYTUM (or-tho-FIGHT-um)

I have rarely seen these plants in collections, yet several are very hardy subjects that can tolerate great abuse and still provide colorful accent. Generally these are small plants, good for window display or desk accent, and require only good light to thrive.

The genus *Orthophytum* consists of about twenty species from Brazil; most are rock-growing plants that bask in direct sun. Plants resemble succulents or cacti because of their spiny and grooved, crinkled, or striated leaves. There are several growth forms, but the rosette type predominates. The diversity of the foliage color makes Orthophytums unusual plants: silver, green, or even vibrant copper. At bloom time the center of the plant turns fiery red, creating a dramatic picture. White flowers are borne on short stems.

If you give Orthophytums sun, they grow into handsome plants, but without adequate light they cannot grow well. I use a potting mix of equal parts gravel, soil, and sand that drains readily. Plants can tolerate coolness if necessary, so a few nights at 50°F (10°C) will not harm them, although an optimum temperature of, say, 65°F (18.3°C) is best. Even in low humidity Orthophytums do well, thus making good houseplants where humidity is lacking. I water my plants sparsely all year. I never flood them, but I never leave them bone dry either. These slow-growing, undemanding plants are worth a try.

O. fosterianum *(fos-ter-ee-**an**-um)* is medium size with 12-inch (30.5-centimeter) apple-green, leathery leaves edged with spines. It bears tufted flower crowns at the leaf axils. A very different bromeliad; the best in the group.

CULTURE

Water: Allow potting medium to dry out between waterings; never soak.

Potting medium: One part gravel, one part soil, one part sand.

Feeding: Occasionally in summer; very weak solution.

Light: Likes sun; do not grow in shade.

Temperature: Prefers coolness, say, about 75°F (23.9°C) during day and 15 degrees cooler at night.

Propagation: From offshoots.

Remarks: Wear gloves when handling spiny leaves. Good houseplants that can survive drought if necessary.

O. navioides *(nay-vee-**oy**-deez)*, with narrow, arching 10-inch (25.4-centimeter) leaves, makes a handsome rosette. It is small and good for a windowsill.

O. saxicola *(sack-see-**kohl**-a)* has small white flowers and broad, pointed 10-inch (25.4-centimeter) leaves armed with spines.

O. vagans *(**vay**-ganz)*, with 10-inch (25.4-centimeter) leaves, is a beauty in bloom; the top leaves of the plant are covered with what looks like red lacquer and small white flowers dot the crown in October.

Orthophytum vagans

PITCAIRNIA (pit-CARE-nee-a)

This is a large genus of bromeliads, but only a few are cultivated in the home. The majority originate in Colombia, Peru, and Brazil. The plants are diversified in habit and growth shape. Some are grasslike, others have succulent-type leaves; many grow in the ground, while some are tree dwellers.

As indoor subjects some of the smaller Pitcairnias do well, but large species are best grown in the temperate garden. The plants are rather scarce and only occasionally listed in suppliers' catalogs.

P. andreana *(an-dree-ay-na)* is perfect for the windowsill. It is 10 inches (25.4 centimeters) high and bears pretty yellow and orange flowers in early spring.

P. corallina *(koh-ra-lee-na)*, the most common cultivated species, grows to 36 inches (91.4 centimeters) with a pendant flower spike and coral-red blossoms. Although difficult to grow, this *Pitcairnia* is a delight in bloom.

P. paniculata *(pa-nik-you-la-ta)*, a 36-inch (91.4-centimeter) fast-growing plant, has an erect spike with brilliant red-and-yellow flowers. A very attractive species that blooms in early October for me.

CULTURE

Water: Keep potting medium fairly moist but never soggy.

Potting medium: Can be grown in packaged soil or use equal parts soil, bark, and sand.

Feeding: Not necessary.

Light: Likes sun but grows in bright light as well.

Temperature: Prefers warmth to 80°F (26.7°C).

Propagation: By division or offshoot.

Remarks: Smaller plants make good indoor subjects; flower crowns decorative.

PORTEA (POR-tee-a)

Growing in Brazil, the Porteas are usually large plants with great beauty. They have spiny leaves and very decorative inflorescence—large and vibrantly colored. They are generally terrestrial and require plenty of sun to bloom. I grow one in a large tub in the garden room where it has done well at a south exposure.

Where an accent plant is needed—and there is space—Porteas make good indoor additions.

P. kermesiana *(ker-mess-ee-na)*, to 26 inches (66 centimeters), has pink flowers on an erect, branched stalk.

P. petropolitana var. extensa *(pet-ro-pol-ee-ta-na, ex-ten-sa)* has large, spreading yellow-green leaves and a 48-inch (1.2-meter) rosette heavily edged with purple spines. The tubular pink, green, and lavender flowers are in a branched raceme; a remarkable combination of colors, followed by blue-lavender berries that last for months.

CULTURE

Water: Keep growing medium evenly moist at all times; large vase of plants should be watered routinely.

Potting medium: Use equal parts soil and bark; bark alone can be used too.

Feeding: Atlas Fish Emulsion a few times a year in summer.

Light: Must have sun to bloom; in bright light will survive but may not bear flowers.

Temperature: Likes warmth to 80°F (26.7°C).

Propagation: By offshoots.

Remarks: Plants are large and need staking or else they have a tendency to tip over.

PUYA (POOY-a)

Puyas are better left outdoors rather than in the home. Most are very large plants with stiff, erect leaves armed with spines. The plants grow in the Andes mountains at about 14,000 feet (4,267 meters) so can take extreme cold weather if necessary (to about 20°F/50.8°C). Indoors they require large containers and a very sunny location to produce their remarkable long spikes of vivid purple-blue flowers—breathtaking to view.

I have never grown the plants indoors but have a friend who has a clump in his yard that blooms every year with 7-foot (2.1-meter) spikes. They are in a sandy rocky soil in full sun and bake during the day and can take temperatures to 30°F (−1.1°C) at night. They flourish. He rarely waters them and allows natural rainfall to do the job. Only mature specimens bloom.

Although most are large plants and do not appeal to me, there are some Puyas that can be grown indoors by the adventurous and you might want to try them.

P. alpestris *(al-pes-tris)* grows to 40 inches (1 meter) with sharp, narrow, recurved leaves. The branched inflorescence is blue and green. A real beauty.

Puya alpestris

CULTURE

Water: Keep somewhat on dry side—overwatering causes harm.

Potting medium: Likes a sandy soil; fir bark and soil satisfactory too.

Feeding: Only occasionally—during summer.

Light: Likes sun or bright light; will not respond in shaded areas.

Temperature: Likes it hot; but needs coolness (55°F/12.8°C) at night.

Propagation: By offshoots in spring.

Remarks: Mostly large plants—not altogether attractive indoors but superb for desert-type gardens.

Group of Puyas

P. berteroniana *(ber-ter-o-**nee**-a-na)* is also large and handsome, to 40 inches (1 meter) with green-and-blue flowers. Outside it will withstand 30°F (−1.1°C) temperatures.

P. chilensis *(chill-**en**-sis)* is 40 inches (1 meter), with a tall, branched flower stalk; this one should be grown outdoors. A really colorful bromeliad in bloom with hundreds of green-and-yellow flowers.

P. mirabilis *(mi-**rab**-i-lis)* grows to 14 inches (35.6 centimeters) with spiny-edged grayish-green flowers.

P. venusta *(ven-**oos**-ta)* to about 10 inches (25.4 centimeters), has shiny green leaves edged with stiff spines. A compact grower.

QUESNELIA (kwes-NAIL-ee-a)

I can still remember the first time I saw *Quesnelia arvensis* (ar-VEN-sis), at a friend's home in Chicago about twenty years ago. I was so stunned by the beauty of the fiery-red inflorescence and handsome foliage that I immediately searched one out and grew it. I think this group of plants has a great future as indoor subjects, and we should see more of them.

Quesnelias are from eastern Brazil, and the thirty species grow in great masses in sandy soils or in the rain forests in shade. The plants are rosette or tubular in shape, with mainly green leaves that have small spines. The inflorescence has bright-rose bracts and pink-and-blue flowers. Most plants are only 30 inches (76.2 centimeters) or so, making them fine for indoor gardens. The leathery texture of the foliage makes it almost impervious to insect attack, and these robust subjects can tolerate abuse and still survive in most situations.

Grow Quesnelias in equal parts of soil and fir bark that drains readily. Plants require a great deal of water; keep vases filled with water at all times. A little liquid feeding (10-10-5) two or three times a year is fine, but do not overdo it. Spray foliage to encourage good humidity. Plants grow well in both shade or sun, but if you want the brilliant floral spikes, sun is absolutely necessary. These bromeliads are outstanding as room accents, and flower crowns stay colorful for many weeks. Plants bear several offshoots. Generally these are difficult plants to find.

Q. arvensis *(ar-***ven***-sis)* is to me the most striking in the genus, with a 30-inch (76.2-centimeter) tubular, vased-shaped plant and crossbanded foliage. The inflorescence is a vivid-red cone stuffed with blue-and-white flowers.

Q. humilis *(***hew***-mil-is)* is delightful, about 10 inches (25.4 centimeters) tall. The light-blue flowers are produced on a short, arching scape in April on my plant; an excellent species.

Q. liboniana *(li-bo-nee-***a***-na)* has a tubular form and grows to 30 inches (76.2 centimeters). The branched inflorescence is red and blue; another handsome houseplant.

Q. marmorata (Aechmea marmorata) *(mar-moor-***a***-ta, Eek-me-a)* makes a 26-inch (66-centimeter) vase-shaped plant with handsome dark-green leaves suffused with maroon; pendant spike with exquisite pink bracts, small blue flowers.

Q. quesneliana *(kwes-nail-ee-***a***-na)* is a large, stiff, vase-shaped species to 20 inches (50.8 centimeters) with a tufted flower crown on an erect stem. A good bromeliad if you have the room for it.

CULTURE

Water: Keep cups filled at all times; allow potting medium to dry out between waterings.

Potting medium: One part soil, one part fir bark.

Feeding: Weak fertilizer once a month in summer.

Light: Can tolerate low light but likes bright exposure.

Temperature: Warm, to 80°F (26.7°C), all year.

Propagation: From offshoots but allow plantlets to get to 6 or 8 inches (15.2 or 20.3 centimeters) before potting up.

Remarks: Very pretty plants, somewhat difficult to find.

RONNBERGIA (ron-BERG-ee-a)

This small genus of terrestrial and epiphytic plants offers some lovely small specimens that are highly desirable for limited spaces. Plants are generally tubular in growth, to about 14 inches (35.6 centimeters), and foliage color is attractive. Coming from the damp forests of Colombia, Panama and Peru,

Quesnelia marmorata

CULTURE

Water: Keep vase of plant filled with water; keep potting medium slightly moist but never soggy.

Potting medium: Use equal parts of soil and fir bark; small pots.

Feeding: No feeding necessary.

Light: Moderate to bright light.

Temperature: Cool preferable, about 55°F to 70°F (12.8°C to 21.1°C).

Propagation: Offshoots root readily.

Remarks: Only a few species available; both small and excellent as table decoration or desk accent.

they grow at altitudes of 6,000 feet (1,829 meters), making them suitable plants for cool apartments and houses (or if you are conserving heat).

Ronnbergias adjust to almost any condition from 55°F to 80°F (12.8°C to 26.7°C) and thrive on bright light. Plants should be potted in equal parts of fir bark and soil and the medium should be kept evenly moist all year. Occasionally spray the growing area to keep humidity at about 40 percent.

These are unassuming plants that deserve more attention and are now available to the general public.

***R. columbiana** (kol-um-bee-a-na)* grows 12 to 16 inches (30.5 to 40.6 centimeters) high; tubular shape, with dark-green leaves shaded dark red. Purple-and-white flowers make this a pretty plant.

***R. morreniana** (mor-ren-ee-a-na),* a 10-inch (25.4-centimeter) tubular plant with few leaves, has colorful, mottled foliage. The slender, erect scape is composed of dense-blue flowers. A handsome species.

STREPTOCALYX (strept-o-CAL-ix)

I like these somewhat large plants; in bloom with red inflorescence they are uncommonly beautiful. If you have a garden room or greenhouse, Streptocalyxes can become a major accent. The genus is closely allied to Aechmeas, and have prickly light-green leaves in rosette growth. Native to Brazil, Ecuador, Peru, and Colombia, plants grow high in trees in hot, humid conditions.

I grow my Streptocalyxes in fir bark in large pots in a sunny exposure where temperatures rarely drop below 60°F (15.6°C). They need a great deal of water and are doused almost daily in summer.

If you are seeking a decorator plant to fill a corner of a room, Streptocalyxes make an effective display in a suitable container. Plants are rarely bothered with insects and grow by themselves—mature specimens, some to 5 feet (1.5 meters) high, are especially effective. Costly, but certainly an exciting challenge for the advanced gardener.

***S. longifolius** (long-gi-fol-ee-us)*, to 36 inches (91.4 centimeters), is a dense rosette of spiny leaves. The ovoid inflorescence is a whitish-rose color on a short scape.

Streptocalyx longifolia

CULTURE

Water: Keep soaked; can take plenty of water but be sure drainage is perfect.

Potting medium: Can be planted in soil or bark alone or equal parts of soil and fir bark.

Feeding: Twice a year with Fish Emulsion.

Light: Likes all the sun it can get.

Temperature: Minumum 60°F to 65°F (15.6°C to 18.3°C). Really likes it hot, preferably about 80°F (26.7°C).

Propagation: By offshoot any time of year.

Remarks: Makes an excellent room plant; superior for vertical decoration. Costly.

S. poeppigii *(pea-pij-ee-eye)* is a 40-inch (1 meter) rosette with narrow, spiny, coppery-red leaves. A handsome foliage plant that bears an erect scape crowned with a cylindrical head of densely set red or rose-purple flowers followed by pink-and-white berries which gradually change to purple and last for months. Leaf color varies according to light exposure. An essential species for the bromeliad enthusiast.

TILLANDSIA (till-AND-see-a)

CULTURE

Water: Plants must never be allowed to be wet; mist frequently in morning but allow to dry off. True air plants.

Potting medium: Some will grow in fir bark but most best grown on tree-fern slabs or pieces of wood.

Feeding: Not needed.

Light: Likes sun but will survive in moderate light if necessary.

Temperature: Can take wide range, from 50°F to 90°F (10°C to 32.2°C).

Propagation: From offshoots; easy.

Remarks: Excellent, offbeat plants that can grow on air if needed with just occasional misting. Must have good air circulation.

This very large group of bromeliads has hundreds of species—more than 500. Plants are variable in shape and size: some are only 1 inch (2.5 centimeters) across, others are 10 feet (3 meters). Tillandsias offer the indoor gardener a host of plants for the house. Most of these epiphytes need no soil; they prefer to grow on tree-fern slabs.

Tillandsias are the predominate bromeliads of North America and are also found in Central America, Argentina, and the West Indies. Most cling to poles or rocks or other plants, many grow in treetops, where they become large colonies. Many Tillandsias are tight rosettes of stiff and leathery or curly leaves, but some plants are urn-shaped, bulbous in growth, or covered with a grayish scale or powder that gives them a frosted look. The soft-leaved species inhabit rain forests; the many silvery types are from dry areas. Tillandsias do not have basic cup growth as do many bromeliads, so they rely on fog and mist for moisture. Thus, indoors plants must be sprayed frequently. Moisture is taken in by the leaves rather than strictly through the roots; usually the root systems are for holding rather than for draining water. Flowers are tubular, and the bracts are bright rose-red. The predominate color of flowers is blue or purple-blue. Most species are small, to about 12 inches (30.5 centimeters), so they are fine for the gardener with limited space.

I grow a few Tillandsias in fir bark in pots, but most are on osmunda wired to trellis work or on cork-bark slabs or tree knolls. They grow readily in this way and plants prefer bright light. Tillandsias tolerate coolness (55°F/12.8°C), but they do best in warmth (75°F/23.9°C and up). Do not feed plants. These plants take care of themselves once they are established, and are slow-growing. Insects rarely attack plants. By all means try a few Tillandsias for your indoor garden.

Note: No specific sizes are given for most Tillandsias because plants vary greatly in size and many have recurving leaves, making it difficult to estimate exact proportions.

T. anceps (an-seps) is a small, stemless species with numerous arching leaves and a large ovoid inflorescence, pale-green or rose with blue petals.

T. brachycollis *(brack-ee-col-lis)* is handsome, with many leaves. At blooming time the center foliage turns from green to coppery-red with purple flowers. Mount this species on a slab of tree fern.

T. bulbosa *(bull-bo-sa)* has a bulbous base with narrow, leathery leaves. The inflorescence is pretty, magenta and white. An oddity best grown on a piece of bark or branch.

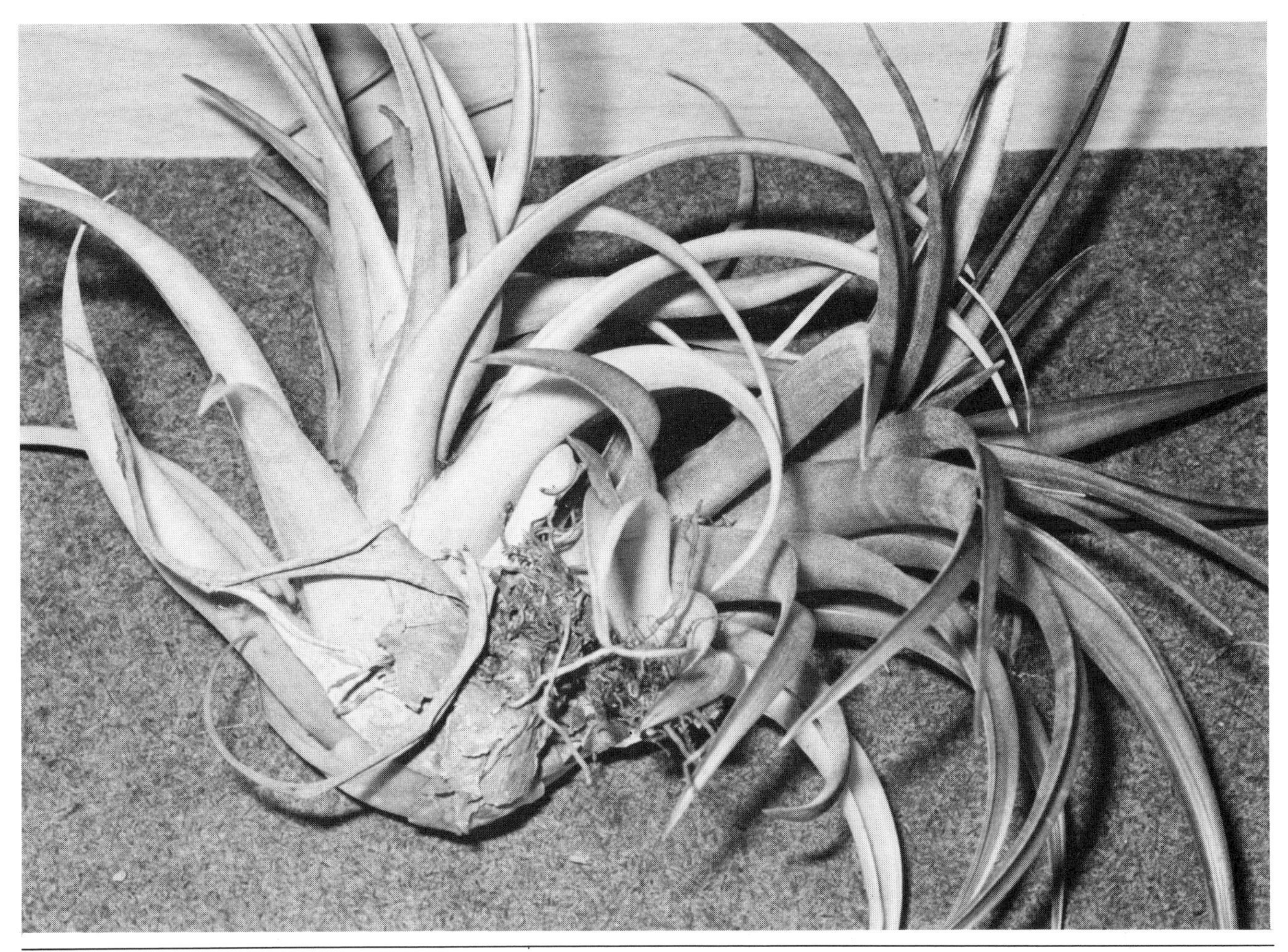

Tillandsia brachybaulos

T. butzii *(**butt**-see-eye)* is small, with thin, twisted, cylindrical leaves that are purple-spotted. The bracts are rose-colored with purple petals and yellow stamens. A pretty species that blooms in spring.

T. capitata 'Giant Orange' *(cap-ee-**ta**-ta)* makes a rosette of leathery grayish-green leaves covered with silver finish, purple-red margins. At bloom time, foliage turns reddish and the inflorescence is a green crown.

T. caput-medusae *(**cap**-it-mi-**dew**-see)* resembles *T. bulbosa*, small with vivid-blue flowers.

T. caulescens *(caw-**les**-sens)* has gray-green spiny leaves, handsome pendant inflorescence with red bracts and yellow flowers.

T. circinnata *(sir-sin-**na**-ta)* is a small rosette with silvery-gray, leathery leaves and small lavender flowers on a flattened spike.

T. concolor *(con-**kol**-or)* has stiff gray leaves; star-shaped rosette. Bracts rose or green on erect spike. Tubular flowers are purple.

T. cyanea *(sigh-**an**-ee-a)*, one of the most popular Tillandias and rightly so, is a regal plant with graceful, arching leaves resembling a palm. From the center of this medium-size species, an erect flower stalk bears a feathery

T. capitata 'Giant Orange'

T. caulescens

T. concolor

T. cyanea

T. dasyliriifolia

pink sword of large, purple-shaded flowers—a stunning bromeliad that needs moisture and humidity to bloom. This is one of the more difficult ones to grow at home but well worth trying.

T. dasyliriifolia *(days-il-eer-if-ol-ee-a)*, a 20-inch (50.8-centimeter) blue-green rosette, has a branching flower stalk that is deep rose, petals whitish-green.

T. fasciculata *(fa-sick-ew-la-ta)* is medium size with blue or purple flowers. Among the many varieties available, leaf and flower color vary somewhat. A good one for the beginner.

T. flexuosa *(flex-ew-o-sa)*, sometimes called ***T. aloifolia*** *(a-loy-fol-ee-a)*, has coppery-green twisted leaves with silver crossbands. It bears red bracts and white flowers.

T. geminiflora *(jem-in-ee-flor-a)*, a dense rosette of green leaves, has a branched inflorescence and is a good plant for a bromeliad tree. An easy species and most decorative.

T. ionantha *(eye-o-nan-tha)* is a dwarf, hardly more than 2 inches (5.1 centimeters) high, but it will astound you at blooming time in spring when all the leaves blush fiery red and tiny, pretty purple flowers appear. Grow it in sun or bright light; it needs little care.

T. fasiculata

T. xerographica

T. juncea *(**jun**-see-a)* is small and pretty, with narrow leaves in a tufted growth and a red flower crown. A favorite of mine.

T. leiboldiana *(lee-bold-ee-**an**-a)* has arching, straplike leaves, wide, gray-green; spotted with maroon, red bracts at bloom time and violet flowers.

T. lindenii *(lin-**den**-ee-eye)* is similar to *T. cyanea*, with long, graceful, tapered, reddish-green leaves. The inflorescence is blue. A real beauty but difficult to bloom.

T. paraenis *(pair-ee-**en**-sis)* is a good windowsill plant. Leaves are grayish-green, the flower bracts pink.

T. punctulata *(punk-tew-**la**-ta)*, with narrow, silver-gray, pointed leaves, bears a heavy inflorescence that is densely set with rose-red bracts and purple-and-white flowers. A good small air plant.

T. streptophylla *(strep-toe-**fill**-a)* is medium size with a bulbous base and grayish-green, curving foliage. The inflorescence is branched, almost the same color as the leaves, with pink bracts. A mature specimen is a handsome sight.

T. stricta *(**strick**-ta)* is a rosette of recurving, leathery leaves covered with silver scales; rose bracts and blue flowers.

T. stricta

T. leiboldiana

T. tricolor *(tree-**kol**-or)*, with grayish-green leaves edged red, is of medium size. The pink-and-red inflorescence is upright and branched, rising well above the plant.

T. xerographica *(zee-ro-**graf**-ee-ka)* is a stiff epiphytic rosette with narrow, concave, recurring silver-gray leaves; branched inflorescence with rose bract leaves; flattened lateral greenish spikes on red stems. The petals are purple.

VRIESEA (VREE-see-a)

CULTURE

Water: Keep cups filled at all times and potting medium evenly moist.

Potting medium: One part fir bark, one part soil.

Feeding: Weak fertilizer to cups and medium once a month (optional).

Light: Moderate to low light; some sun fine.

Temperature: Wide range from 55°F to 90°F (12.8°C to 32.2°C).

Propagation: From offshoots, but these are scarce. Also from seed in seven years.

Remarks: Likes humid conditions, say, 50 percent. Very decorative plants.

Years ago one plant in this group brought Vrieseas recognition: the Flaming Sword, *V. splendens* (SPLEN-denz), which opened the door for many of the other Vrieseas. In Europe, Vrieseas have long been good houseplants.

There are about 225 species of Vrieseas inhabiting a large region from Mexico, Central America, and the West Indies to Peru, Bolivia, Paraguay, northern Argentina, and Brazil. These plants grow in rain forests but are also found at altitudes of over 8,000 feet (2,438 meters). Vrieseas are mainly epiphytic, inhabiting tree branches in areas of very good air circulation. They need dappled light. A few of the larger plants grow as terrestrials, but this is rare. Most are vase-shaped rosettes similar in appearance to Aechmeas and Guzmanias. The majority are medium sized, about 36 inches (91.4 centimeters) across, and have smooth leaves. Foliage is plain or apple-green to an almost purple-green or barred, striped, banded, or spotted in contrasting colors. When inflorescence develops, the plants are a veritable rainbow. The bracts are highly colored—red, green, or purple—and last for weeks. The small flowers are generally in shades of yellow.

Grow Vrieseas in a bright place—direct sun is not needed. At a west window my plants do fine. I grow them in fir bark and soil; some are suspended on rafts. I frequently water them in hot weather and always keep the vases filled with water. I spray plants frequently in the warm months with tepid water. I do not feed the plants. Although Vrieseas can tolerate some cold, they prefer warmth (75°F/23.9°C) and seem immune to any insect pests or disease. For the gardener who "can't grow anything," Vrieseas are the answer because they grow on their own. There are many fine hybrids available.

V. barilletii *(bar-il-**let**-ee-eye)*, with a 24-inch (61-centimeter) rosette of shiny green leaves, bears an erect yellow-and-red inflorescence that is colorful from December to April. Not as showy as most species but very dependable.

V. bituminosa *(bi-tew-min-**o**-sa)*, to 40 inches (1 meter) across, has a large, stocky rosette with broad blue-green leaves purple-tipped; inflorescence on erect spike with scattered bracts on either side of stalk; flowers are yellow.

V. carinata *(kar-in-**a**-ta)*, to 10 inches (25.4 centimeters) across, is good for a windowsill. It has light-green leaves and a flat, sword-shaped red-and-yellow flower head.

V. c. aurea *(**or**-ee-a)* grows to 14 inches (35.6 centimeters) with apple-green leaves; flower spike bright yellow with trace of red.

Vriesea (unidentified)

V. bituminosa

V. fenestralis

V. c. 'Mariae,' the Painted Feather plant, is a somewhat larger hybrid. It holds color all winter.

V. fenestralis *(fen-ee-**stral**-is)*, a regal 40-inch (1-meter) rosette with green leaves, delicately figured darker green and purple lined, is big and bushy but worth the space it takes; a most desirable bromeliad with yellow flowers. Grow this one warm.

V. friburgensis *(free-bur-**gen**-sis)*, is a 10-inch (25.4-centimeter) rosette of curved shiny-green leaves; bears a branching inflorescence with green bracts and yellow flowers.

V. gigantea *(jy-**gan**-tee-a)* is a 30-inch (76.2-centimeter) rosette of blue-green leaves mottled with yellow and green crosslines; tall, branched inflorescence with green bracts and yellow flowers. Also called ***V. tesselata*** *(tess-ee-**la**-ta)*.

V. heliconoides *(hell-ee-kon-**oy**-deez)* shouts with color, a 14-inch (35.6-

centimeter) rosette. The flower spike is bright red, edged with chartreuse, and the green leaves are suffused with red—a spectacular plant in bloom.

V. hieroglyphica *(high-er-o-**glif**-ee-ka)*, with a rosette to 40 inches (1 meter), is light green, crossbanded dark green, and purple-brown beneath. Makes a fine decorative plant. The yellow flowers appear in spring on a tall, branched spike. Grow this species at a northern exposure with good humidity.

V. imperialis *(im-peer-ee-**a**-lis)*, a 60-inch (1.5-meter) rosette, has green or dark wine-red leaves crowned with a tall branch of red-and-yellow inflorescence. A majestic bromeliad that requires space.

V. 'Kitteliana,' a hybrid growing to 16 inches (40.6 centimeters), has dark olive-green leaves spotted with burgundy, scapes brownish-red, yellow flowers.

V. malzinei *(mal-**zin**-ee-eye)*, with a compact rosette of claret-colored leaves shaded green, rarely grows over 16 inches (40.6 centimeters) and is fine for the windowsill. Early in spring the cylindrical flower head has yellow bracts with green margins. Very easy to grow.

V. perfecta *(per-**fect**-a)* is a hybrid, a popular species but not half as showy as others I grow. It grows to 30 inches, has apple-green foliage and produces a yellow-and-red flower spike.

V. friburgensis

V. heliconoides

V. hieroglyphica

V. petropolitana *(pet-ro-pol-ee-****tan****-a)* is a charmer. Only 10 inches (25.4 centimeters) high, it bears brilliant orange-and-yellow flowers in early spring.

V. platynema variegata *(plat-ee-****nee****-ma var-ee-****ga****-ta)* is a 40-inch (1-meter) rosette of blue-green leaves; the inflorescence is featherlike with purple bracts and greenish-white flowers.

V. regina *(re-****jy****-na)* forms a 48-inch (1.2-meter) rosette of handsome green, waxy leaves speckled at the base; tall inflorescence is branched with rosy bracts.

V. 'Rubin,' a recent hybrid, has an open rosette to 14 inches (35.6 centimeters), green, glossy, fiery red bracts, and a flower crown with yellow petals.

V. schwackeana *(shwak-ee-****an****-a)* grows to 20 inches (50.8 centimeters) across with green leaves and a strong scape of four to six dark-red and yellow ovoid heads. Blooming time is spring.

V. 'Kitteliana'

V. splendens (splen-denz), another good *Vriesea,* grows about 12 inches (30.5 centimeters) and is a perfect houseplant. The green foliage is mahogany striped and the thin, thrusting spring and summer inflorescence is orange-colored.

V. s. 'Meyer's Favorite' grows to 48 inches across, a glowing rosette of apple-green leaves blotched dark green; tall, fiery crown.

V. s. mortefontanensis *(mor-tee-fon-tan-**en**-sis)* grows to 24 inches (61 centimeters) across, with handsome green foliage banded with brown; tall, typical "flaming sword" flower spike.

V. vagans *(**vay**-ganz)* grows in clusters with individual rosettes about 8 inches (20.3 centimeters) in diameter; leaves are green, blackish at base, orange petals. Can tolerate coolness.

WITTROCKIA *(wit-ROCK-ee-a)*

A small group of plants, Wittrockias offer a few medium-size species that are very suitable to indoor culture. The most popular species grown, *W. superba* (soo-PER-ba), has stiff spiny leaves and the inflorescence is sunk low in the cup of the plants similar to Nidulariums. The few other species have thin leaves with marginal spines.

V. platynema variegata

V. schwackeana

V. splendens var. mortefanensis

Wittrockia superba

CULTURE

Water: Keep leaf cup filled with water; potting medium barely moist.

Potting medium: Soil or bark-and-soil mixture; be sure drainage is good.

Feeding: Not necessary.

Light: Will tolerate shade but bright light provides better leaf coloring.

Temperature: Average home temperatures (60°F to 80°F/15.6°C to 26.7°C).

Propagation: By offshoots.

Remarks: Good room plants; easy to grow and very colorful.

Plants are from the mountains of Brazil and like a somewhat sunny place. I grow mine in fir bark and soil with excellent drainage and they do very well. If necessary, Wittrockias can tolerate some coolness (to 50°F/10°C) with no ill effects.

Not a spectacular genus of plants but worthwhile for those looking for something different.

W. amazonica *(am-a-**zon**-ee-ca)* is a 28-inch (71.1-centimeter) rosette, the flower head sunk in the center of the rosette.

W. superba *(soo-**per**-ba)*, to 24 inches (61 centimeters), has dark-green spiny-edged foliage handsomely tipped red. The inflorescence is slightly raised.

Seven: Hybrids, Bigenerics, and Collector's Items

In the odds and ends—the collector's or rare plants, the bigenerics—we find tomorrow's new plants. Hybridization in bromeliads is only in its infancy, and the rare species that is the collector's plant today becomes everyone's plant tomorrow.

COLLECTOR'S PLANTS

Many bromeliad enthusiasts delight in having a rare species—not generally available—growing in their greenhouse or garden room. Perhaps it is a status symbol or simply the desire to see a plant grow and bloom that few people have seen. Whether it is for prestige or adventure, the one-of-a-kind species is always sought after. Suppliers have these plants in limited quantity and they are always expensive—you should expect to pay at least $30 to $100. If you wait a few years until the plants become plentiful, you can have them at moderate cost.

In this group of rare plants are species such as *Aechmea bracteata variegata*, prized for its exquisite leaf coloring. Some of the *A. chantinii* hybrids are also highly desirable: 'Green Goddess,' for example and 'Ash Blonde.' There are in fact dozens of handsome bromeliads you can grow that few people have and that are available from mail-order suppliers. Some are true species from nature while others are hybrids—the best crossed with the best within a given genus.

***Neomea* 'San Diego'**

BIGENERICS

A hybrid is a cross between the same plants in a genus, but bigenerics are accomplished by mating a species from one genus with a species from another genus. One example is *Aechularium* 'Santa Cruz,' a cross between *Aechmea* and *Nidularium. Neomea,* a popular cross between *Neoregelia* and *Aechmea,* is another example of a very beautiful plant, and Quesnelias crossed with *Aechmeas* give us our *Quesmias.* A cross between *Orthophytum* and *Cryptanthus* has produced *Orthotanthus* 'What,' a lovely white-and-green striped plant of small proportions, superior to the usual *Orthophytum* species.

OTHER BROMELIADS

I have written here about the bromeliads I have grown through the years or ones I have had firsthand knowledge of, but there are other groups of bromeliads—lesser known—that I have not included mainly because they are infrequently cultivated and I have rarely seen them at growers.

The plants are mainly from dry regions of South America and in most cases there may be only one species known in each genus. Here is a brief rundown on these plants:

Androlepis *(an-dro-**lep**-is):* Only one species in this genus: *A. skinnerii* (SKIN-er-ee-eye), a very large plant with bright-rose leaves armed with spines. Bracts and flowers are both yellow.

Brocchinia *(bro-**kin**-ee-a):* These are mainly terrestrial plants that grow very large in a semirosette form. Seldom seen in cultivation.

Deuteroconnia *(dew-ter-o-**co**-nee-a):* These plants closely resemble the Puyas and have the same stiff, spined leaves in a rosette shape.

Fosterella *(fos-ter-**el**-a):* Small terrestrial plants, this genus recognized in 1960; plants generally have grayish-green leaves in a flat rosette; flowers white.

Greigia *(**greeg**-ee-a):* There are about twenty species in this genus and the plants are dense rosettes of dark-green leaves edged with spines. Growing at high altitudes, few of these plants survive average home conditions.

Navia *(**nav**-ee-a):* Many species in the genus but few if any have found their way into cultivation; plants vary in size from small to large.

Part 3/CHOOSING BROMELIADS

Easy-to-Grow Bromeliads

Abromeitella brevifolia
Acanthostachys strobilacea
Aechmea fasciata
Aechmea fulgens discolor
Aechmea mertensii
Aechmea ornata
Aechmea racinae
Billbergia euphemiae
Billbergia meyeri
Billbergia 'Muriel Waterman'
Billbergia nutans
Billbergia splendens
Catopsis floribunda
Cryptanthus (all species)
Guzmania lingulata
Guzmania vittata
Guzmania zahnii
Hechtia glomerata
Hechtia montana
Neoregelia carolinae
Neoregelia cruenta
Neoregelia 'Marmorata'
Neoregelia spectabilis
Nidularium regelioides
Orthophytum navioides
Orthophytum saxicola
Pitcairnia andreana
Quesnelia arvensis
Ronnbergia morreniana
Tillandsia caulescens
Tillandsia concolor
Tillandsia cyanea
Tillandsia ionantha
Tillandsia tricolor
Vriesea barilletii
Vriesea carinata
Vriesea heliconoides
Vriesea malzinei
Vreisea petropolitana
Vriesea schwackeana
Vriesea splendens

Bromeliads for Direct Light

Abromeitella brevifolia
Abromeitella chlorantha
Aechmea angustifolia
Aechmea chantinii
Aechmea luddemanniana
Aechmea mexicana
Aechmea mooreana
Aechmea orlandiana
Aechmea nudicaulis
Aechmea 'Red Wing'
Aechmea 'Spring Beauty'
Ananas bracteatus
Ananas comosus
Ananas nanus
Araeococcus flagellifolius
Araeococcus pectinatus
Billbergia euphemiae
Billbergia 'Fantasia'

Billbergia meyeri
Billbergia pyramidalis var. striata
Billbergia splendens
Bromelia humilis
Dyckia (most species)
Hechtia (most species)
Hohenbergia ridleyi
Hohenbergia stellata
Neoregalia ampullacea
Neoregalia 'Bonfire'
Neoregalia compacta
Neoregalia concentrica
Neoregalia cruenta
Neoregalia johannis
Neoregalia 'Painted Lady'
Puya (most species)
Streptocalyx longifolius
Tillandsia anceps
Tillandsia brachycollis
Tillandsia butzii
Tillandsia flexuosa
Tillandsia geminiflora

Bromeliads for Diffused Light

Aechmes caudata variegata
Aechmea fasciata
Aechmea chantinii
Aechmea ramosa
Aechmea weilbachii
Billbergia nutans
Billbergia pyramidalis var. concolor
Billbergia sanderiana
Billbergia venezueleana
Cryptanthus (most species)
Neoregelia marmorata
Orthophytum (most species)
Portea petropolitana
Quesnelia humilis
Tillandsia bulbosa
Tillandsia fasiculata
Tillandsia streptophylla
Vriesea heliconoides
Vriesea hieroglyphica
Vriesea imperialis
Wittrockia superba

Bromeliads for Semi-Shade

Aechmea filicaulis
Aechmea racinae
Canistrum cyathiforme
Canistrum lindenii albo marginata
Guzmania monostachia
Guzmania musaica
Guzmania zahnii
Neoregelia carolinae
Neoregelia carolinae 'Meyendorfii'
Nidularium innocentii lindenii
Quesnelia arvensis
Ronnbergia columbiana
Ronnbergia morreniana
Vriesea pertropolitana
Vriesea schwackeana
Vriesea splendens

Bromeliads at a Glance

Name	Size	Flower	Leaf	Growth	Exposure
ABROMEITELLA					
# *brevifolia*	small	white	gray-green	rosette	full sun
# *chlorantha*	small	white	silver-green	mats	full sun
ACANTHOSTACHYS					
strobilacea	small	orange	reddish brown	pendant	partial sun
AECHMEA					
• *angustifolia*	medium	yellow	greenish brown	tubular	partial sun
bracteata	large	red yellow	apple green	rosette	bright light
brevicaulis	medium	orange yellow	light green	tubular	partial sun
calyculata	medium	yellow	green	vase	diffused light
caudata var. variegata	medium	orange	green, ivory	rosette	diffused light
caudata albo marginata	medium	orange	green, ivory	rosette	diffused light
chantinii	large	yellow, red	green, silver-banded	vase	partial sun
chantinii 'Burgundy'	large	yellow, red	plum	vase	partial sun
chantinii 'Silver Ghost'	large	yellow	green	vase	partial sun
chantinii x A. ramosa	medium	red yellow	olive-green	vase	partial sun
distichantha var. schlumbergeri	medium	violet	gray-green	tubular	partial sun
fasciata	medium	blue	olive-green, silver-banded	vase	diffused light
fasciata 'Silver King'	medium	pink, blue	green striped	vase	diffused light
fasciata variegata	medium	pink, blue	yellow, green striped	vase	diffused light
filicaulis	small	white	glossy green	vase	diffused light
• 'Foster's Favorite'	large	blue	wine-red	vase	diffused light
• *fulgens var. discolor*	large	violet	olive-green, purple	vase	diffused light
• *luddemanniana*	large	red and blue	green-mottled dark-green	vase	full sun
• *mertensii*	small	red and yellow	green	tubular	diffused light
'Meteor'	large	red	pale green	tubular	diffused light
mooreana	medium	rose and orange	bronze-green	rosette	partial sun

= terrestrial (unmarked species considered epiphytes although some adapt to terrestrial growth)
• = colorful berries

full sun = 6 hours
partial sun = 4 hours
diffused (bright) light = 2 hours
semishade = light, but no sun

small = to 12 inches (30.5 centimeters)
medium = to 36 inches (91.4 centimeters)
large = to 60 inches (1.5 meters)

Name	Size	Flower	Leaf	Growth	Exposure
nudicaulis variegata	medium	red and yellow	gray-green, silver-banded	tubular	partial sun
orlandiana	medium	white, yellow	green, brown-banded	vase	full sun
ornata	medium	pink	gray-green	vase	diffused light
* *penduliflora*	medium	yellow	reddish brown	rosette	diffused light
* *pubescens*	medium	ivory	gray-green, green	tubular	diffused light
quesnelia marmorata x A. chantinii	medium	red, green	gray-green	tubular	diffused light
* *racinae*	small	red, yellow, black	green	vase	shade
'Rajah'	small	orange, yellow	pale-green	tubular	diffused light
* *ramosa*	medium	yellow	apple green	vase	diffused light
recurvata var. ortgiesii	small	pink	dark green	tubular	diffused light
* 'Red Wing'	medium	straw	copper	rosette	partial sun
'Spring Beauty'	medium	rose, yellow	olive	rosette	partial sun
* *tillandsioides*	small	yellow-green, yellow	gray-green	tubular	diffused light
tillandsioides lutea	small	orange, yellow	gray-green	rosette	diffused light
weilbachii	medium	red, lavender	green	vase	diffused light
zebrina	large	yellow	olive-green	vase	partial sun
ANANAS					
# *bracteatus*	large	lavender	green	rosette	full sun
# *bracteatus variegatus*	large	lavender	green	rosette	full sun
# *comosus*	large	purple	gray-green	rosette	full sun
nanus	small	purple	dark green	rosette	full sun
ARAEOCOCCUS					
* *flagellifolius*	medium	pink	reddish brown	bottle	partial sun
pectinatus	small	red	reddish brown	bottle	partial sun
BILLBERGIA					
amoena	medium	green, blue	green	tubular	diffused light
amoena var. viridis	medium	green	green, red, cream, rose	tubular	diffused light
brasiliensis	medium	rose, blue	silver-banded	tubular	partial sun
distachia	small	green, blue	pinkish brown	tubular	partial sun
elegans	large	rose, green	gray-green	tubular	partial sun
euphemiae	small	pink, blue	gray-green	tubular	partial sun
'Fantasia'	medium	blue	green, spotted cream and pink	tubular	partial sun
horrida	medium	green, blue	brown, silver-banded	tubular	diffused light
leptopoda	small	green, blue	green, spotted cream color	tubular	diffused light
lietzei	small	cerise	green	tubular	diffused light

Name	Size	Flower	Leaf	Growth	Exposure
meyeri	medium	blue, green	gray-green, silver-banded	tubular	partial sun
'Muriel Waterman'	small	pink, blue	plum, banded	tubular	partial sun
nutans	medium	blue, green	dark green	pendant	diffused light
porteana	medium	green	gray-green	tubular	diffused light
pyramidalis var. concolor	medium	pink	golden green	tubular	diffused light
pyramidalis var. striata	large	pink	green, yellow	tubular	partial sun
sanderiana	medium	green, blue	green	rosette	diffused light
splendens	medium	pink, blue	pale green	tubular	partial sun
venezueleana	large	purple	brown, silver-banded	tubular	diffused light
vittata	medium	red, violet	olive green	tubular	partial sun
zebrina	large	golden yellow	gray green, silver-flecked	tubular	diffused light
BROMELIA					
# *balansae*	large	rose, white	dark green	rosette	full sun
# *humilis*	medium	pink, white	green	rosette	full sun
serra variegata	medium	rose	green, white, pink	rosette	full sun
CANISTRUM					
cyathiforme	medium	yellow	green, brown	rosette	semishade
fosterianum	medium	white	green, mottled brown	tubular	diffused light
# *lindenii*	medium	white	green	rosette	diffused light
# *lindenii albo marginata*	small	whitish green	yellow-green	rosette	semishade
lindenii var. roseum	medium	white	dark green	vase	diffused light
CATOPSIS					
berteroniana	medium	white	apple green	bottle	diffused light
# *floribunda*	small	white	apple green	bottle	diffused light
# *morreniana*	small	yellow-white	apple green	vase	diffused light
nutans	small	yellow	green	vase	diffused light
sessiliflora zebrina	small	white	dark green	tubular	partial sun
CRYPTANTHUS					
# *acaulis*	small	white	green	rosette	diffused light or semishade
# 'Aloha'	small	white	multicolored	rosette	diffused light
# *beuckeri*	small	white	greenish cream	rosette	diffused light or semishade
# *bivittatus*	small	white	pinkish brown, silver-green	rosette	diffused light or semishade
# *bromelioides*	small	white	greenish brown	rosette	diffused light or semishade
# *bromelioides var. tricolor*	small	white	green, pink, cream	rosette	diffused light or semishade
# 'Bueno Funcion'	small	white	gray-green	rosette	diffused light
# *fosterianus*	medium	white	dark brown, crossbanded	rosette	diffused light or semishade

Name	Size	Flower	Leaf	Growth	Exposure
# *fosteriana* 'Elaine'	small	white	bronze-brown	rosette	diffused light
'It'	small	white	rose, green	rosette	diffused light
# 'Koko'	small	white	green	rosette	diffused light
# 'Minibel'	small	white	rose	rosette	diffused light
# *tricolor*	small	white	brown, silver	rosette	diffused light
# *zonatus*	small	white	brownish green to copper, crossbanded	rosette	diffused light or semishade
DYCKIA					
# *brevifolia*	small	orange	dark green	rosette	partial sun
# *fosteriana*	medium	orange	silver-green	rosette	partial sun
# *frigida*	large	orange	waxy green	rosette	partial sun
# 'Lad Cutak'	small	orange	bronze-green	rosette	full sun
# *leptostachya*	medium	orange	reddish brown	rosette	partial sun
# *marnier lapostellei*	small	orange	gray-green	rosette	partial sun
# *rariflora*	small	orange	silver-gray	rosette	partial sun
FASCICULARIA					
pitcairnifolia	large	blue	dark green	rosette	full sun
GUZMANIA					
berteroniana	medium	yellow	wine red, or green	rosette	diffused light or semishade
lindenii	medium	green, white	green, mottled	rosette	partial sun
lingulata	medium	orange-red, white	apple green	rosette	diffused light or semishade
lingulata var. major	medium	orange, white	green	rosette	partial sun
lingulata var. minor	medium	orange	green	rosette	diffused light or semishade
'Magnifica'	medium	white	shiny green	rosette	diffused light or semishade
# 'Meyer's Favorite'	small	red	green, red	rosette	diffused light
# 'Minnie Exodus'	small	red, yellow	pale green	rosette	partial sun
monostachia	medium	red, brown, white	dark green	rosette	diffused light or semishade
musaica	medium	golden white	green, dark green, red-brown	rosette	diffused light or semishade
'Orangeade'	medium	orange, yellow	green	rosette	partial sun
'Symphonie'	medium	crimson, yellow	copper	rosette	partial sun
vittata	medium	white	light green, chocolate-banded	rosette	diffused light or semishade
zahnii	small	white	light green, penciled maroon-red	rosette	diffused light or semishade
zahnii variegata	small	red, white	green, yellow	rosette	partial sun

Name	Size	Flower	Leaf	Growth	Exposure
HECHTIA					
# *argentea*	small	orange	glossy green	rosette	full sun
# *glomerata*	medium	white	glossy green	rosette	full sun
# *montana*	small	greenish white	dark green	rosette	full sun
# *rosea*	small	pink	brown-red	rosette	full sun
# *texensis*	small	pink	brownish green	rosette	full sun
HOHENBERGIA					
ridleyi	large	lavender	golden green	vase	full sun
stellata	large	purple	golden green	vase	full sun
NEOMEA					
'San Diego'	large	brown, green	green	rosette	partial sun
NEOREGELIA					
ampullacea	small	white, blue	glossy green, brown cross-bands	rosette	partial sun
'Bonfire'	medium	blue	plum	rosette	partial sun
carolinae	medium	purple, white	dark green	rosette	diffused light or semishade
carolinae 'Meyendorfii'	medium	lilac	olive, copper	rosette	semishade
carolinae var. tricolor	medium	purple, white	green, white stripes	rosette	diffused light or semishade
compacta	medium	lilac	green	rosette	partial sun
concentrica	medium	blue	pale green	rosette	partial sun
cruenta	medium	blue	golden green	rosette	partial sun
johannis	small	lavender, blue	green	rosette	partial sun
'Marmorata'	medium	white	green, red-marbled	rosette	partial sun
'Painted Lady'	medium	violet	dark green, red	rosette	partial sun
'Purple Passion'	small	pink	purple-violet	rosette	partial sun
'Red Knight'	small	violet	green, maroon	rosette	partial sun
spectabilis	medium	blue	olive-green	rosette	partial sun
zonata	small	blue	greenish purple	rosette	partial sun
NIDULARIUM					
billbergioides	small	white	green	rosette	diffused light
billbergioides var. citrinum	medium	white	green	rosette	diffused light
fulgens	medium	blue	yellow, green, spotted dark green	rosette	diffused light
innocentii	medium	white	purple	rosette	diffused light
innocentii lindenii	medium	white	green, white	rosette	diffused light
innocentii var. lineatum	medium	white	green, ivory-striped	rosette	diffused light
innocentii var. wittmackianum	medium	white	green	rosette	diffused light
procerum	large	orange, red	yellow-green	rosette	diffused light
regelioides	medium	red	dark green	rosette	diffused light

Name	Size	Flower	Leaf	Growth	Exposure
ORTHOPHYTUM					
# *fosterianum*	medium	white	apple green	branched rosette	diffused light
# *navioides*	small	white	green	rosette	diffused light
# *saxicola*	small	white	green	rosette	diffused light
# *vagans*	small	white	metallic green	rosette	diffused light
ORTHOTANTHUS					
'What'	small	white	green, white	rosette	partial sun
PITCAIRNIA					
# *andreana*	small	yellow, orange	gray-green	branched	diffused light
# *corallina*	medium	red	gray-green	stalk	diffused light
# *paniculata*	medium	red	green	stalk	diffused light
PORTEA					
kermesiana	medium	pink	green	vase	diffused light
* *petropolitana var. extensa*	large	pink, green, lavender	apple green	rosette	diffused light
PUYA					
# *alpestris*	large	blue, green	green	dense rosette	full sun
# *berteroniana*	large	blue	green	dense rosette	full sun
# *chilensis*	large	green	green	dense rosette	full sun
# *mirabilis*	small	green	gray-green	dense rosette	full sun
# *venusta*	small	green	green	dense rosette	full sun
QUESNELIA					
arvensis	large	blue and white	dark green, banded	rosette	diffused light or semishade
humilis	small	blue	gray-green	tubular	diffused light
liboniana	medium	red, blue	dark green	tubular	diffused light
marmorata	medium	pink, blue	dark green, maroon	tubular	diffused light
quesneliana	large	blue, white	green	rosette	diffused light
RONNBERGIA					
columbiana	small	purple, white	greenish brown	tubular	semishade
morreniana	small	blue	bright green, spotted dark green	stalk	semishade
STREPTOCALYX					
longifolius	medium	white	dark green	rosette	partial sun
* *poeppigii*	large	rose-purple	pinkish green	rosette	partial sun
TILLANDSIA					
anceps	small	blue	dull green	dense rosette	partial sun
brachycollis	small	purple	reddish brown, green	dense rosette	partial sun
bulbosa	small	purple, white	silver-green	bottle	diffused sun
butzii	small	purple, yellow	green, spotted purple	bottle	partial sun

Name	Size	Flower	Leaf	Growth	Exposure
capitata 'Giant Orange'	small	orange	gray-green	rosette	partial sun
caput-medusae	small	blue	greenish	bottle	partial sun
caulescens	small	red, yellow	gray-green	rosette	partial sun
circinnata	small	lavender	silver, green	rosette	partial sun
concolor	small	rose-purple	gray-green	rosette	partial sun
cyanea	medium	violet blue	dark	rosette	diffused light
dasyliriifolia	small	blue	gray-green	rosette	partial sun
fasciculata	medium	blue	gray-green	rosette	diffused light
flexuosa	small	white	gray-green, silver-banded	bottle	partial sun
geminiflora	small	yellow, lavender	purplish gray	rosette	partial sun
ionantha	small	purple	silver green	rosette	partial sun
juncea	small	blue, purple	olive-green	rosette	partial sun
leiboldiana	small	violet	gray-green	rosette	partial sun
lindenii	medium	blue	dark green	rosette	diffused light
paraensis	small	red, yellow	dull green	bottle	partial sun
punctulata	small	purple, white	gray-green	rosette	partial sun
streptophylla	medium	lilac	silver-green	bottle	diffused light
stricta	small	blue	gray-green	bottle	diffused light
tricolor	medium	violet, white	dark green	rosette	partial sun
xerographica	medium	rose, purple	silver-gray	rosette	partial sun
VRIESEA					
barilletii	small	yellow	green	rosette	diffused light or semishade
bituminosa	medium	yellow	green	rosette	partial sun
carinata	small	red, yellow	green	rosette	diffused light or semishade
carinata aurea	small	yellow, red	green	rosette	partial sun
carinata 'Mariae'	medium	yellow, green	green	rosette	diffused light or semishade
fenestralis	large	yellow	light green, lined dark green and purple	rosette	diffused light or semishade
friburgensis	small	green, yellow	green	rosette	partial sun
gigantea	medium	green, yellow	green	rosette	partial sun
heliconoides	medium	white	dark green suffused with red	rosette	diffused light or semishade
hieroglyphica	large	yellow	green zig-zagged with purple	rosette	diffused light or semishade
imperialis	large	yellow	wine-red	rosette	diffused light or semishade
'Kitteliana'	small	yellow	olive green	rosette	partial sun

Name	Size	Flower	Leaf	Growth	Exposure
malzinei	small	yellow, green	claret color	rosette	diffused light or semishade
perfecta	medium	yellow	green	rosette	diffused light or semishade
petropolitana	small	yellow, orange	green	rosette	diffused light or semishade
platynema variegata	medium	purple, greenish-white	blue-green, striated	rosette	diffused light
regina	large	rose	green	rosette	partial sun
'Rubin'	small	red	apple green	rosette	partial sun
schwackeana	medium	yellow	dark green, spotted purple	rosette	diffused light or semishade
splendens	small	yellow	green, purple bands	rosette	diffused light or semishade
splendens 'Meyer's Favorite'	large	red	apple green	rosette	partial sun
splendens mortefontanensis	medium	red	green	rosette	partial sun
vagans	small	orange	green	rosette	partial sun
WITTROCKIA					
amazonica	medium	white	dark green with purple	vase	diffused light
superba	medium	blue	dark green, tipped red	vase	diffused light

Glossary

Aphid Plant lice.
Axil Where two leaves meet.
Banded Marked with bars or lines of color.
Bicolored Two-colored.
Bigeneric A cross between species of different genera.
Botrytis Disease of plants.
Bowl Center of bromeliad plant.
Bract A modified leaf.
Calyx The outermost case of a flower.
Concolor One color.
Cultivar A horticulturally or agriculturally derived variety of a plant, as distinguished from a natural variety.
Cup A water reservoir that is created by the closeness of the leaves in the center of a plant.
Discolor Of two, or of different, colors.
Epiphyte An air plant; a plant that grows on trees or other plants but is not parasitic.
Genera More than one genus.
Genus A group of related species.
Habitat Particular area in which a plant grows.
Hybrid A cross between two plants.
Inflorescence The part of the plant that holds the flower. Also referred to as a flower.
Kiki An offshoot or offset; small plant.
Lanceolate Like a lance; a narrow leaf.
Margin The edge and the area immediately adjacent to it, such as a border.
Mealybug Small white cottony plant insect.
Mother plant The original or parent plant that is producing or has produced offsets.
Offset An offshoot; a plant growing close to the base of the mother plant.
Osmunda Roots of tree fern fiber.
Panicle A loose, branching cluster of flowers.
Pendant Hanging.
Petiole The stalk or stem of a leaf.
Propagation To increase or multiply by methods of reproduction.
Pup Same as *Offset*.
Raceme An elongated cluster with stalked flowers.
Recurve To bend or curve backwards or downwards.
Rosette A circular arrangement of clustering leaves.
Scales Minute, flat organs on many bromeliads that absorb water and nutrients.
Scape The stem of the inflorescence.
Sepal One of the leaves of a calyx.
Serrated Toothed.
Species Subdivision of a genus.
Spike A compact, elongated inflorescence.
Stolon A shoot that bends to the ground; partial root.
Striated Having thin lines or bands, especially those that are parallel and close together.
Style The stem part of a pistil.
Sucker Same as *Offset*.
Systemic Chemical that makes plant parts poisonous to insects.
Tank Same as *Cup*.
Taxonomist Authority who names plants.
Terete Slenderly tapering.
Terminal Leaves at the end of the stem.

Terrestrial Growing in soil.
Tomentose Covered with dense, short, or matted hairs.
Tubular Tubular in shape as in a vase.
Variety A naturally occurring or selectively bred variation of a species.
Vase Same as *Cup*.
Vermiculite A sterile potting medium.
x Indicates a cross between the two named parents.
Xerophytic A plant that grows in and is adapted to an environment that is deficient in moisture.

Bibliography

Baker, J. G. *Handbook of the Bromeliaceae*. London, 1889.

Bromeliad Society Bulletin, vols. I—XXVIII. Los Angeles: *Journal of the American Bromeliad Society*.

Duval, Léon. *Les Bromeliacées*. Librairie Agricole, de La Maison Rustique, 26 Rue Jacob, Paris, France.

Foster, Mulford B. *Brazil, Orchid of the Tropics*. Ronald Press, New York, 1945.

Gilmartin, A. J. *The Bromeliaceae of Ecuador*. 1972.

———. *Las Bromeliacias de Honduras*, Ceiba 11(2), 1965.

Kramer, Jack. *Bromeliads: The Colorful House Plants*. New York: Van Nostrand, 1965.

Pallida, Victoria. *Bromeliads*. New York: Crown Publishers, 1978.

Rauh, Werner. *Bromeliads*. Poole, Dorset, England: Blandford Press, 1979.

Richter, W. *Anzucht und Kultur der Bromelien mit besonderer Berücksichtigung den für den Handel wichtigsten Arten. Grundlagen und Fortschritte im Gartenund Weinbau*, part 76. Stuttgart, 1950 (out of print).

———. *Zimmerpflanzen von heute und morgen: Bromeliaceen*. Dresden, 1962.

Smith, L. B. "The Bromeliaceae of Peru," in F. Macbride, ed., *Flora of Peru*, part 1, no. 3. Chicago: Field Museum of Natural History, 1936.

———. *Studies in the Bromeliacea*. No. I-XVII in *Contributions from the Gray Herbarium of Harvard University*, in vols. 1930–1946 (Parts I-XIV) and *Contributions from the U.S.—National Herbarium*, vol. 29, 1949–1954 (Parts XV-XVIII), reprinted.

———. "The Bromeliacea of Brazil," in *Smithsonian Miscellaneous Collections*, vol. 126, 1955. Reprinted as *Contribution from Reed Herbarium*, no. XXVI, 1977.

Wilson, Catherine and Bob. *Bromeliads in Cultivation*. Coconut Grove, Florida: Hurricane House Publishers, 1963.

Appendix A: Note on Bromeliad Societies

Most cities have their own chapter of a bromeliad society. At their meetings you can meet other people interested in the plants and can exchange ideas about cultivation, new species, supplying and the like. Many cities also hold annual bromeliad shows. To find out about membership fees and other information, check the proper listings in your telephone book.

For general information on bromeliad societies, write to The Bromeliad Society, Inc. P.O. Box 41261, Los Angeles, CA 90041.

Appendix B: Suppliers

The following plant suppliers are those I have personally purchased plants from, so that I can vouch for their reputation in the industry and the quality of their plants. In addition, there are dozens of other mail-order suppliers in various parts of the United States that include bromeliads in their catalogs. Check the Yellow Pages of local telephone books. The omission of these suppliers in no way denotes their stock is inferior—only that I have never received any plants from them and cannot make any comments.

Alberts & Merkel Boynton Beach, FL 33435	*Excellent color catalog (25¢); good selection of plants and hybrids.*
California Jungle Gardens 11977 San Vicente Blvd. Los Angeles, CA 90049	*Bromeliad list.*
Exotic Bromeliads 744 E. Valencia St. Lakeland, FL 33801	*No list.*
Fantastic Gardens 9550 South West 67th Ave. Miami, FL 33156	*Large selection of top-quality plants; list available.*
Holmes Nurseries P.O. Box 17157 Tampa, FL 33607	*Good stock; brochure available.*
Lee Moore P.O. Box 504B Kendall, FL 33156	*Bromeliad list (50¢).*

North Jersey Bromeliads 15 Douglas Dr. Hillsdale, NJ 07642	*No list.*
Oak Hill Gardens P.O. Box 25 Rte. 2, Bimmie Road Dundee, IL 60118	*Excellent stock, price list available.*
Seaborn Del Dios Nursery Rte. 3, Box 455 Escondido, CA 92025	*Catalog $1.*
Earl J. Small P.O. Box 11207 St. Petersburg, FL 33733	*Catalog.*

Index

Page references in boldface refer to black and white illustrations.

Abromeitella, 79–81
cultural directions for, 79
A. *brevifolia*, 79, **80,** 157, 159
A. *chlorantha*, **80,** 81, 157, 159
Acanthostachys, **33,** 81
A. *strobilacea*, 81, 157, 159
see also color insert
Aechmea, 20, 28, **29, 34, 36, 69,** 81–91, 110, 131, 140, 152
cultural directions for, 81–82
flowers of, 13, **18,** 81, 82
growth pattern of, 13, 93
leaves of, 8, 81
natural habitat of, 8, 81
propagation of, 53, 55, 56, 81
see also color insert
A. *angustifolia*, **24,** 82, **82,** 157, 159
see also color insert
A. *bracteata*, 82–83, 159
A. *b. variegata*, 151
A. *brevicolis*, 83, 159
see also color insert
A. *calyculata*, 53, 83, **83,** 159
A. *caudata albo marginata*, 83, 159
A. *c. var. variegata*, 83, 158, 159
A. *chantinii*, 16, 82, 83, **84,** 157, 158, 159
A. *c.* 'Ash Blonde,' 151
A. *c.* 'Burgundy,' **11,** 84, 159
see also color insert
A. *c.* 'Green Goddess,' 151
A. *c.* 'Silver Ghost,' 84, 159
see also color insert
A. *chantinii x A. ramosa*, 159
see also color insert
A. *distichantha var. schlumbergeri*, 84, 159
A. *fasciata*, 4, 8, 16, 20–21, 24, **24, 68, 69,** 84, 157, 158, 159
see also color insert
A. *f.* 'Silver King,' 84, 159
A. *f. variegata*, **20,** 85, 159
see also color insert
A. *filicaulis*, 85, 158, 159
A. 'Foster's Favorite,' 85, 159
A. *fulgens var. discolor*, **19,** 85, **85,** 157, 159
see also color insert
A. *luddemanniana*, 82, 86, 157, 159
A. *marmorata*, 129
A. *mertensii*, 86, **86, 87,** 157, 159
see also color insert
A. 'Meteor,' 4, 86, 159
see also color insert
A. *mexicana*, 157
A. *mooreana*, 86, 157, 159
see also color insert
A. *nudicaulis variegata*, 87, 157, 160
see also color insert
A. *orlandiana*, 88, 157, 160
A. *ornata*, 88, 157, 160
A. *penduliflora*, 88, 160
A. *pubescens*, 88, 160
A. *quesnelia marmorata x A. chantinii*, **88,** 89, 160
A. *racinae*, 81, 89, 157, 158, 160
A. 'Rajah,' 89, 160
see also color insert
A. *ramosa*, 89, 158, 160
A. *recurvata var. ortgiesii*, 89, **89,** 160
A. 'Red Wing,' 90, **90,** 157, 160
see also color insert
A. 'Spring Beauty,' 90, 157, 160
see also color insert
A. *tillandsioides*, 90, 160
A. *t. lutea*, 91, 160
see also color insert
A. *weilbachii*, 91, **91,** 158, 160
A. *zebrina*, 15, 91, 98, 160
see also color insert
Aechularium 'Santa Cruz,' 152
Aineb, 63
air circulation, 8, 49, 75
alcohol, as insect remedy, 61
altitude, plant distribution and, 6–8
Ananas, 92
A. *bracteatus*, 92, 157, 160

A. b. variegatus, 92, 160
A. comosus, 92, 157, 160
see also color insert
A. nanus, 92, 157, 160
Androlepis, 153
A. skinnerii, 153
aphids, 20, 59, 60, **60,** 61
apples, in forcing of flowers, 56, **58**
Araeococcus, 92–93
A. flagellifolius, 92–93, 157, 160
A. pectinatus, 93, 157, 160

bacteria, 63
Benomyl, 63
berries, 8, **18**
bigenerics, 152
Billbergia, 28, **34,** 81, 93–98
cultural directions for, 93
flowers of, 13, 16, 93
growth pattern of, **11,** 13, 93
natural habitat of, 8, 93
propagation of, 53, 55, 56
B. amoena, 93, 160
B. a. var. viridis, 93–94, 160
B. brasiliensis, 94, 160
see also color insert
B. distachia, 94, 160
B. elegans, 94, 160
see also color insert
B. euphemiae, 94, **94,** 157, 160
B. 'Fantasia,' 94–95, 157, 160
B. horrida, 95, 160
B. leptopoda, 95, 160
B. lietzei, 95, 160
see also color insert
B. meyeri, 95, 157, 158, 161
B. 'Muriel Waterman,' 95, **95,** 157, 161
B. nutans, **17,** 93, 95–97, 157, 158, 161
see also color insert
B. porteana, 97, 161
B. pyramidalis, 16
B. p. var. concolon, 97–98, 158, 161
see also color insert
B. p. striata, **96,** 158, 161
B. sanderiana, 98, 158, 161
B. splendens, **96,** 98, 157, 158, 161
B. venezueleana, 98, 158, 161
B. vittata, 98, 161
see also color insert
B. zebrina, 16, 93, **97,** 98, 161
Black Leaf, 40, 62
botrytis, 63
Brocchinia, 153
Bromelia, 98–99
B. balansae, 98, 99, 161
B. humilis, 98, 99, 158, 161
B. serra variegata, 99, 161
see also color insert

bromeliads:
anatomy of, **4,** 8–13
buying of, 22–25, 27
care for, 18–20, 45–58 (*see also specific genuses*)
cataloguing of, 26
collecting of, 25–27
cost of, 24–25
decorative foliage of, 14, **14,** 15, 20
easy-to-grow, 157
first, suggestions for, 20–22
at a glance, 159–166
growth patterns of, **11,** 13 (*see also specific genuses*)
at home, 28–44
life cycle of, 53
moved indoors, 64, 67, 68–70
naming of, 3–4, 23
natural habitats of, 5–8, **5, 6, 7** (*see also specific genuses*)
new, tips for, 25, 27
outdoor cultivation of, 64–71
packaging of, 26
problems of, 59–63
bromeliad "trees," 36–38, **37**
"host" selected for, 38
Bromelioideae, 3

Canistrum, **13,** 99–100
cultural directions for, 99
propagation of, 53, 99
C. cyathiforme, 99, 158, 161
see also color insert
C. fosterianum, 99, 161
C. lindenii, 100, **100,** 161
see also color insert
C. l. albo marginata, 100, 158, 161
C. l. var. roseum, 100, **101,** 161
Captan, 63
Catopsis, 100–102
cultural directions for, 101
growth pattern of, **11**
C. berteroniana, 101, 161
C. floribunda, 102, 157, 161
C. morreniana, 102, 161
C. nutans, 102, 161
C. sessiliflora zebrina, 102, **102,** 161
chlorine, in tap water, 48
clay pots, 29, **31,** 50
collector's plants, 151
containers, 29–33, **30, 31,** 64
drainage holes in, 31, 50
glazed pottery as, 31
hanging, 34, 38
plastic vs. terra-cotta, 29
preparation of, 50
wire and wood, 31, **32**

Cryptanthus, 22, **34, 37,** 50, 103–107, 152
cultural directions for, 103–104
growth pattern of, **11,** 13, 102
leaves of, 15, 102–103
light conditions for, 104, 157, 158
see also color insert
C. acaulis, 104–105, 161
C. 'Aloha,' **104,** 105, 161
C. beuckeri, 105, 161
C. bivittatus, 105, 161
C. bromelioides, 105, 161
C. b. var. tricolor, 105, 161
C. 'Bueno Funcion,' **105,** 107, 161
C. fosterianus, 107, 161
C. f. 'Elaine,' 107, 162
see also color insert
C. 'It,' 107, 162
see also color insert
C. 'Koko,' **106,** 107, 162
C. 'Minibel,' **106,** 107, 162
see also color insert
C. tricolor, 107, 162
C. zonatus, 107, 162
cups, *see* vases

Deuteroconnia, 153
dew, 8
Diazinon, 62
diseases, 48, 49, 50, 63, 75
division, propagation by, 55, **55**
driftwood, bromeliads grown on, **35, 36**
see also color insert
Dyckia, 43, 79, 107–108, 158
cultural directions for, 107
flowers of, 14
growth pattern of, 13
D. brevifolia, 107, 162
D. fosteriana, 107, 162
D. frigida, 107, 162
D. 'Lad Cutak,' **108,** 108, 182
D. leptostachya, 108, 162
D. marnier lapostollei, 108, **109,** 162
D. rariflora, 108, 162
endangered species, 27
epiphytes, bromeliads as, 5, 20, 29, 66
ethylene gas, in forcing of flowers, 56

fall, bromeliad care in, 52–53
Fascicularia, 109
F. pitcairnifolia, 109, 162
see also color insert
feeding, 48–49, 75
year-round schedule for, 52, 53

Ferbam, 63
fir bark, as planting medium, 45, 48, 50, **51**

flowers, 8, 16, 50, 53
berries and, **18**
forcing of, 56, **58**
forms of, **10, 12, 15, 16, 17, 19, 21**
light needed for, 14
stalks of, 8
times for, 13
see also specific genuses
fluorescent lamps, 41, 42, **42**
incandescent lamps used with, 43, 44
types of, 43–44
fog, 8
Fosterella, 153
fungicides, 63
fungus, 48, 49, 63, 75

gardens, 64–66, **65, 66, 68**
bromeliads grown year-round in, 64–65
genuses, naming of, 3–4
greenhouses, 64, 72–75, **72, 73, 74**
drawbacks of, 74–75
Greigia, 153
grooming, 49–50
Guzmania, 20, **30, 66, 69,** 110–115, 140
cultural directions for, 110
flowers of, 13, 16, 110
growth pattern of, **11,** 93, 110
leaves of, 8, 110
mounting of, 36, 38
natural habitat of, 8, 110
propagation of, 56, 110
G. berteroniana, 110, 162
G. lindenii, 15, 110, 162
G. lingulata, 13, 16, 21, **22, 23,** 24, 110, 113, 157, 162
G. l. major, 110, **111,** 162
see also color insert
G. l. var. minor, 110, 162
see also color insert
G. 'Magnifica,' 110–112, 162
see also color insert
G. 'Meyer's Favorite,' **111,** 112, 162
G. 'Minnie Exodus,' 112–113, **112,** 162
see also color insert
G. monostachia, 13, 16, 113, **113,** 158, 162
G. musaica, 113, 158, 162
G. 'Orangeade,' 113, 162
see also color insert
G. 'Symphonie,' 113–114, **114,** 162
see also color insert
G. vittata, **35,** 114, 157, 162
G. zahnii, 15, 16, 114–115, 157, 158, 162
see also color insert
G. z. variegata, 115, **115,** 162

handpicking method, as insect remedy, 61
hanging containers, 34, 38
Hechtia, 43, 115–117, 158
cultural directions for, 116
flowers of, 14, 116
leaves of, 8, 115
H. argentea, 116, 163
H. glomerata, 116, **116,** 157, 163
H. montana, 117, **117,** 157, 163
H. rosea, 117, 163
H. texensis, 117, 163
Hohenbergia, **30,** 117–118
cultural directions for, 118
H. ridleyi, 118, 158, 163
H. stellata, 118, 158, 163
see also color insert
humidity, 33, 49, 63
in greenhouses, 74, 75
year-round schedule for, 52–53
humus, in potting soil, 48
hybrids, 4, 20, 151, 152

importing of plants, 25–27
collection techniques for, 25–26
through foreign suppliers, 27
permits for, 25, 27
incandescent lamps, 42, 43, 44
heat given off by, 43, 44
inflorescence, *see* flowers
insecticides, 26, 61–62
six rules for, 62
trade or brand names of, 62
insect pests, 8, 20, 59–62, **60, 61,** 75
eggs of, 25, 61
inspections for, 59–60
in newly bought plants, 23, 25
old-fashioned remedies for, 61
in outdoor cultivation, 64, 68–70
soaking of pots and, 25, 60, 70
interior decoration, bromeliads as, 28–29, **29, 30, 33**
Isotex, 62

Karanthane, 63
kikis (offshoots), 53, 54

leafmold, as planting material, 50
leaf-shining preparations, 25
leaves, 8, 20
as decorative foliage, 14, **14,** 15, 20
markings of, **9, 11**
see also specific genuses
light, 3, 8, 29, 49, 52, 74
blooming and, 14
diffused, bromeliads for, 158
direct, bromeliads for, 157–158
for newly bought plants, 25, 27

light *(cont.)*
semi-shade, bromeliads for, 158
as used by plants, 42
in window arrangements, 36
year-round schedule for, 52, 53
light, artificial, 41–44, **41,** 56
correct amount of, 43, 44
as dramatic accent, 41, 44
fluorescent, 41, 42, **42,** 43–44
incandescent, 42, 43, 44
manufacturers of, 43–44

mail-order suppliers, 23–24
Malathion, 62
mealybugs, 20, 59, 60, 61, **61**
metal trays, 32–33
Meta-Systox, 62
Mexico, four climatic areas in, 6
mildew, 48, 63
misting, 20, 26, 48, 52
mounted bromeliads, 36–40, **35, 36**
assembling of, 38, **40**
in gardens, 64–65
on patios and terraces, 66–68
selection of materials for, 36
as "trees," 36–38, **37**
watering of, 38–39

Navia, 153
Neomea, 152
N. 'San Diego,' **152,** 163
see also color insert
Neoregelia, 22, **35, 65, 66, 73,** 99, 118–122, **119,** 152
cultural directions for, 118–119
flowers of, 8, 13, **17,** 118
growth pattern of, **11,** 13, 118
leaves of, 15, 118
natural habitat of, 8, 118
propagation of, 53, 56, 118
N. ampullacea, 119, 158, 163
N. 'Bonfire,' 119, 158, 163
N. carolinae, 28, 119, 157, 158, 163
see also color insert
N. c. 'Meyendorfii,' 119, 158, 163
N. c. 'Meyendorfii' *variegata, see color insert*
N. c. var. tricolor, 22, 119, 163
see also color insert
N. compacta, 120, 158, 163
see also color insert
N. concentrica, 120, **120,** 158, 163
N. cruenta, 120, **121,** 157, 158, 163
see also color insert
N. johannis, 120, 158, 163
N. 'Marmorata,' 120, 157, 158, 163
see also color insert
N. 'Painted Lady,' **121,** 122, 158, 163

N. 'Purple Passion,' 122, 163
see also color insert
N. 'Red Knight,' 122, 163
see also color insert
N. spectabilis, 122, 157, 163
see also color insert
N. zonata, 122, **122,** 163
Nidularium, 99, 122–124, 147, 152
cultural directions for, 123
flowers of, 8, 13, 123
growth pattern of, 13, 123
leaves of, 15, 122, 123
natural habitat of, 8, 122
propagation of, 53, 56, 123
N. billbergioides, 123, 163
N. b. var. citrinum, 123, 163
N. fulgens, 123–124, **123,** 163
N. innocentii, 124, 163
N. i. lindenii, 124, 158, 163
see also color insert
N. i. var. lineatum, 22, 124, 163
see also color insert
N. i. var. wittmackianum, 124, 163
N. procerum, 124, 163
N. regelioides, 124, 157, 163
nurseries:
mail-order, 23–24
outdoor, 23

offshoots (kikis), 53, **54**
orchids, **6, 7, 74**
Orthophytum, 124–125, 152, 158
cultural directions for, 124
O. fosterianum, 124, 164
O. navioides, 15, 125, 157, 164
see also color insert
O. saxicola, 125, 157, 164
O. vagans, 125, **125,** 164
Orthotanthus, 152
O. 'What,' 164
see also color insert
osmunda (tree-fern root)
in mounting of bromeliads, 36, 38, 64–65, 68
as planting medium, 45, **46, 47,** 50
as starting medium, 56

patios, 66–68, **67, 69**
Peru, five climatic areas in, 8
pets, bromeliads eaten by, 33
Pitcairnia, 126
P. andreana, 126, 157, 164
P. corallina, 126, 164
P. paniculata, 126, 164
Pitcairnioideae, 3
planting mediums, 45–48, **46, 47**
function of, 45
storage of, 48

Plant Quarantine Division, U.S., Permit Unit of, 25
plant shops, 22–23
plastic containers, 29, 32
poles, suspension, 34, **38, 39**
Portea, 126
P. kermesiana, 126, 164
P. petropolitana var. extensa, 126, 158, 164
see also color insert
potting and repotting, 20, 22, 27, 50–52, **51**
plants prepared for, 50–52
see also containers; planting mediums
propagation, 53–56
by division, 55,**55**
with offshoots, 53, **54**
with seeds, 56, **57**
see also specific genuses
Puya, 126–128, **128,** 153, 158
cultural directions for, 126–127
flowers of, 14, 126
P. alpestris, **7,** 126, **126,** 164
P. berteroniana, 128, 164
P. chilensis, 128, 164
P. mirabilis, 128, 164
P. venusta, 128, 164
Pyrethrum, 62

Quesmias, 152
Quesnelia, 129, 152
cultural directions for, 129
flowers of, 16, 129
propagation of, 53, 129
Q. arvensis, 129, 157, 158, 164
Q. humilis, 129, 158, 164
Q. liboniana, 129, 164
Q. marmorata, 129, **130,** 164
see also color insert
Q. quesneliana, 129, 164

red spider mites, 59, 60
repotting, *see* potting and repotting
Ronnbergia, 129–130
R. columbiana, 130, 158, 164
R. morreniana, 130, 157, 158, 164
rot, 50, 63
Rotenone, 62
runners (stolons), 53

sand, in potting soil, 48
scale (insects), 59, 60, 61
scales (on leaves), 8
seeds, 56, **57**
Sevin, 62
shards, in potting, 50, 51
soap, as insect remedy, 61
soil, as planting medium, 45–48, 50, **51**
species, naming of, 3–4
sphagnum moss, as starting medium, 56
spring, bromeliad care in, 52
starting mediums, 56
stolons (runners), 53
Streptocalyx, 130–132
cultural directions for, 131
flowers of, 16, 130
S. longifolius, 131, **131,** 158, 164
S. poeppigii, 132, 164
subfamilies, 3
Sulfur, 63
summer, bromeliàd care in, 52
systemics, 62

temperature, 8, 33, 49, 52, 64
in greenhouses, 75
year-round schedule for, 52–53
terraces, 66–68, **67, 69**
terrariums, **37,** 56, **57**
Tillandsia, **72,** 132–140
cultural directions for, 132
flowers of, 13, 14, 16, 132
mounting of, 36, **37,** 38, **47,** 132
potting of, 20, 132
propagation of, 53, 56
see also color insert
T. anceps, 132, 158, 164
T. brachybaulos, **133**
T. brachycollis, 132, 158, 164
T. bulbosa, 132, 158, 164
T. butzii, 133, 158, 164
T. capitata 'Giant Orange,' 133, **134,** 165
T. caput-medusae, 133, 165
T. caulescens, 16, 133, **134,** 157, 165
see also color insert
T. circinnata, 133, 165
see also color insert
T. concolor, 133, **135,** 157, 165
T. cyanea, 8, 21, 133, **136,** 157, 165
see also color insert
T. dasyliriifolia, 136, 137, 165
T. fasciculata, 16, 137, **137,** 158, 165
see also color insert
T. flexuosa, 137, 158, 165
T. geminiflora, 137, 158, 165
T. ionantha, **36,** 48, 137, 157, 165
see also color insert
T. juncea, 138, 165
T. leiboldiana, 138, **139,** 165
T. lindenii, 138, 165
T. paraenis, 138, 165
T. punctulata, 138, 165
T. streptophylla, 138, 158, 165
T. stricta, 138, **139,** 165
T. tricolor, 140, 157, 165

T. xerographica, **138,** 140, 165
Tillandsioideae, 3
tobacco, as insect remedy, 60
tree-fern root, *see* osmunda
trimming, 49–50

varieties, naming of, 4
vases (cups), 8–13
  water kept in, 13, 20, 50, 56, 75
vermiculite, as starting medium, 56
Vriesea, 20, **34,** 140–147, **141**
  cultural directions for, 140
  flowers of, 14, 140
  growth pattern of, 13, 140
  leaves of, 8, **14,** 140
  propagation of, 53
V. barilletii, 140, 157, 165
  see also color insert
V. bituminosa, 140, **141,** 165
V. carinata, 21, **24,** 140, 157, 165
V. c. aurea, 140, 165
  see also color insert
V. c. 'Mariae,' 142, 165
V. fenestralis, 15, 142, **142,** 165
V. friburgensis, 142, **143,** 165
V. gigantea, 142, 165
V. heliconoides, 142–143, **144,** 157, 158, 165
V. hieroglyphica, 15, 143, **145,** 158, 165
V. imperialis, 143, 158, 165
V. 'Kitteliana,' 143, **146,** 165
V. malzinei, 143, 157, 166
V. perfecta, 143, 166
V. petropolitana, 145, 157, 158, 166
  see also color insert
V. platynema variegata, 145, **147,** 166
V. regina, 145, 166
  see also color insert
V. 'Rubin,' 145, 166
  see also color insert
V. schwackeana, 145, **148,** 157, 158, 166
V. splendens, 21–22, **67,** 140, 146, 157, 158, 166
V. s. 'Meyer's Favorite,' 146, 166
  see also color insert
V. s. mortefontanensis, 146, **148,** 166
V. tesselata, 142
V. vagans, 146–147, 166

washing of bromeliads, 59
watering, 3, 48
  in greenhouses, 74, 75
watering *(cont.)*
  misting in, 20, 26, 48, 52
  of mounted bromeliads, 38–39
  of outdoor bromeliads, 65–66
  after potting, 52
  vase filled in, 13, 20, 50, 56, 75
  year-round schedule for, 52, 53
  water spray method, as insect remedy, 61, 70
window arrangements, 28, 33–36
  base materials for, 34, **34, 35**
  light requirements in, 36
  plants suspended in, 34, **38, 39**
  trays for, 32–33
winter, bromeliad care in, 53
wiping method, as insect remedy, 61
wire baskets, as containers, 31
Wittrockia, 147–150
  cultural directions for, 150
W. amazonica, 150, 166
W. superba, 147, **149,** 150, 158, 166
wood containers, 31, **32**

year-round schedule for bromeliads, 52–53